Our Molecular Nature

Our Molecular Nature

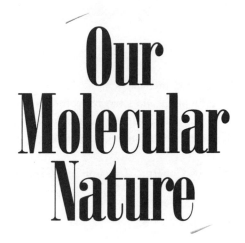

The Body's Motors, Machines and Messages

David S. Goodsell

COPERNICUS
An Imprint of Springer-Verlag

Published in the United States by Copernicus, an imprint of Springer-Verlag New York, Inc.

Copernicus
Springer-Verlag New York, Inc.
175 Fifth Avenue
New York, NY 10010
USA

Library of Congress Cataloging-in-Publication Data
Goodsell, David S.
    Molecular nature / the body's motors, machines and
    messages / David S. Goodsell.
        p.   cm.
    Includes bibliographical references and index.
    ISBN 0-387-94498-2 (hardcover : alk. paper)
    1. Biochemistry.   2. Molecular biology.   3. Biomolecules.
    I. Title.
        [DNLM: 1. Molecular biology.   2. Biochemistry.   3. Amino Acids.
    QH 506 G589m 1996]
    QP514.2G66   1996
    574.8'8—dc20   95-46846

Acquiring Editor: Robert Garber.
Manufactured in the United States of America.
Printed on acid-free paper.

9   8   7   6   5   4   3   2

ISBN 0-387-94498-2   SPIN 10567779

"Nature" is what we see —
The Hill — the Afternoon —
Squirrel — Eclipse — the Bumble bee —
Nay — Nature is Heaven —
Nature is what we hear —
The Bobolink — the Sea —
Thunder — the Cricket —
Nay — Nature is Harmony —
Nature is what we know —
Yet have no art to say —
So impotent Our Wisdom is
To her Simplicity.

Emily Dickinson

Preface

Busily at work within each of us is the most complex array of machinery known on the Earth. This book is an exploration of these machines: the molecules that make us what we are. These tiny molecules, millions of times smaller than machines in our familiar world, capture energy and resources from food; they shape our bodies and power our motion; they protect us and repair damage; and they orchestrate our inner worlds of sense, emotion, and thought. These remarkable machines are our most valuable personal possessions, inherited from our parents at birth and used every second of our lives.

This book is necessarily eclectic. Tens of thousands of different molecules guide our daily life processes—only the 150 or so that have been the most extensively studied are included here. For each, I have focused on molecules as they relate to our familiar lives: how we might recognize the individual actions of alcohol dehydrogenase or opsin, and what we would feel if they were not present. An enormous body of literature is available for each of the molecules pictured and described in this book. I have merely touched the surface of each, providing an illustration and a few interesting details. Readers interested in more detailed study of the structure of each molecule are refered to the individual sources, included in the Appendix. Readers interested in the intricate biochemistry of these molecules are refered to the list of general references as an entrance to further study.

I thank Arthur J. Olson and T. J. O'Donnell for insightful comments on the manuscript and Teresa A. Larsen and William P. Grimm for their constructive critique of the illustrations.

Contents

1 The Molecular World
Looking at the Molecular World   3
The Names of Molecules   8
Molecular Building Blocks   8

2 Building Molecules
Enzymes   20
Information   36

3 Powering the Body
Digestion   56
Chemical Energy   67

4 Form and Motion
Cell Form   82
Body Form   95
Molecular Motors   105

5 Dangers and Defenses
Toxins and Venoms   109
Detoxification   116
Immunity   124
Healing   131

6 Molecules and the Mind
Hormones   141
Sense   154
Thought   159

Epilogue: Molecules and Medicine
165

Further Reading
167

Glossary
169

Sources of Macromolecular Structures
173

Index
177

1
The Molecular World

An invisible world courses through every inch of your body. This is the world of *molecules*. Your molecules are tiny machines—millions of times smaller than machines in our familiar world—each performing one microscopic task. Each drop of your blood contains hundreds of different kinds of molecules: some transport food and air, others carry messages, others stand ready to repair an injury. In your eyes, a glittering solution of molecules refracts and focuses an image of light. A subtle change in the shape of another molecule captures this image like a sheet of film. At this moment, in your arms and hands, legions of tiny molecular motors are laboring to hold this book and to turn each page. Indeed, these words you are reading are recognized, sorted, and understood in a flurry of electrochemical impulses, organized by ranks of molecular switches. These everyday tasks—breathing, seeing, moving, and thinking—may be traced to the combined action of invisible molecules. Individually, each of your molecules is a delicate instrument. Together, they endow you with life.

The molecules of living things, including our own molecules, are unique among the molecules of the Earth. These tiny molecular messengers, engines, and machines are built according to a specific plan, each to perform a specific task. Molecules formed by physical processes never show the complexity and design of your own molecules. Earthly molecules are ruled by simple laws: quartz will always grow in angular crystals, later to be ground into sand; water will always freeze to form hexagonal plates, occasionally of incredible, but still strictly hexagonal, beauty. We, on the other hand, build thousands of different molecules, starting only with the limited collection of building blocks available in the diet. When hungry, we build tough, bean-shaped chymotrypsin (page 21) to digest our food. When ill, we build Y-shaped antibodies (page 124) to fight the attacking bacteria or viruses. With every thought, we build tiny neurotransmitters (page 161) to communicate one portion of the idea. The instructions to build our personal molecules, finely honed over the course of evolution, are stored in every cell, available on demand.

There is nothing abstract or mysterious about the science of molecular biology. Most of the molecules pictured in this book are performing their tiny jobs at this very moment, somewhere inside your body. As we explore the many shapes and sizes of these molecules, imagine each performing its job in some corner of your own body. The following chapter—Building Molecules—explores the molecules that direct the construction of new molecules. Imagine these molecules building towers of hair on your scalp and tough fingernails on your hands. Think of them healing small cuts, reconnecting torn tissues, and building new skin and blood vessels. Ponder the fact that, in the first few weeks of your life, your entire body was built by these molecules from two tiny cells and their 46 strands of DNA, all smaller than a single grain of salt. The third chapter—Powering the Body—deals with molecules that harness chemical energy, the energy to power and control your body. Think of these molecules in your stomach, digesting food into usable bits and pieces. Remember the flush of energy rushing through your body after you eat molecules of sugar, and think of these molecules spreading through your blood. The fourth chapter—Form and Motion—describes the most familiar of your personal molecules. The color of your hair, the texture of your skin, and the strength of your arms are due to the properties of these molecules. The fifth chapter—Dangers and Defenses—deals with the molecules of detoxification, immunity, and repair. Think back to the last time you were ill and imagine the internal warfare that occurred throughout your body. Think of how quickly small injuries are blocked up with a blood clot, with no conscious help from yourself. The sixth chapter—Molecules and the Mind—deals with the molecules of communication. These molecules are perhaps the most difficult to imagine. They carry messages throughout your body. Some are experienced as sense, some as emotion, others as thought and memory, and some as imagination itself.

Looking at the Molecular World

Our picture of the molecular world is almost entirely a mental picture. Except in special cases, individual molecules are completely invisible. Under the light microscope, individual molecules are too tiny to be resolved, but cells may be easily seen. Small, free-living cells like yeast or bacteria, a few thousandths of a millimeter across, are transparent spheres rushing and jostling on the microscope slide. Human cells are larger, perhaps a hundredth of a millimeter across. Individual cells may be scraped from inside the cheek or seen in a drop of blood. Under the microscope, they appear as translucent blocks of gelatin, but looked at carefully, they are a flurry of motion inside. By staining cells with colored dyes, convoluted internal compartments and textures may be observed. A large compartment, the nucleus, is seen toward the center. If we look very closely, we may see many small compartments, some rounded, some long and stringy, in the surrounding areas. However, the molecules forming these compartments, and the molecules held within them, are 1,000 times smaller and cannot be resolved in a conventional light microscope. Individual molecules require more sophisticated apparatus to probe their secrets.

Electron microscopes are often useful for looking at individual molecules, particularly very large molecules. Instead of focusing a beam of light through the specimen, as in a light microscope, the electron microscope uses a beam of electrons. Instead of using glass lenses, an electron microscope focuses the beam with magnetic lenses similar to those in a television, producing a ghostly picture on a phosphor screen. Biological molecules pose a major problem for the electron microscope: they are composed of light atoms, like carbon and nitrogen, that are not heavy enough to stop many electrons, and therefore they appear almost completely transparent. So, biological molecules are often coated with metal atoms, such as uranium or gold, in order to be opaque enough to create an image. Electron micrographs, then, are often pictures of the *outsides* of molecules. What we see is the shape of this thin metal plating. In most cases individual atoms cannot be discerned, only the general shape and size of the molecule. Small molecules may appear simply as tiny, round blobs. To improve this image, we need to turn to x-rays.

X-rays are fine enough to resolve individual atoms in a molecule, so an x-ray microscope would be ideal. In a microscope, a beam of light or electrons hits

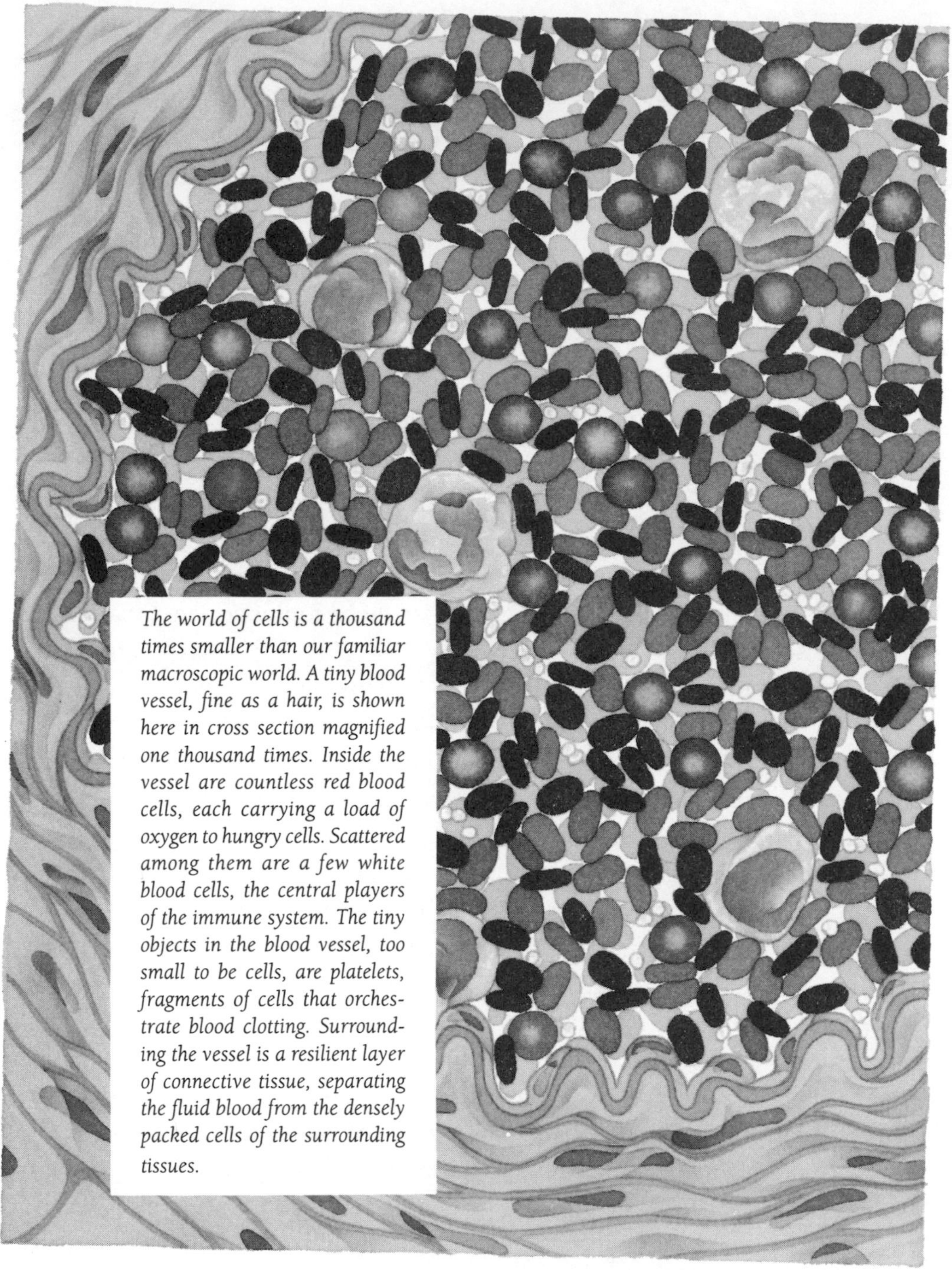

The world of cells is a thousand times smaller than our familiar macroscopic world. A tiny blood vessel, fine as a hair, is shown here in cross section magnified one thousand times. Inside the vessel are countless red blood cells, each carrying a load of oxygen to hungry cells. Scattered among them are a few white blood cells, the central players of the immune system. The tiny objects in the blood vessel, too small to be cells, are platelets, fragments of cells that orchestrate blood clotting. Surrounding the vessel is a resilient layer of connective tissue, separating the fluid blood from the densely packed cells of the surrounding tissues.

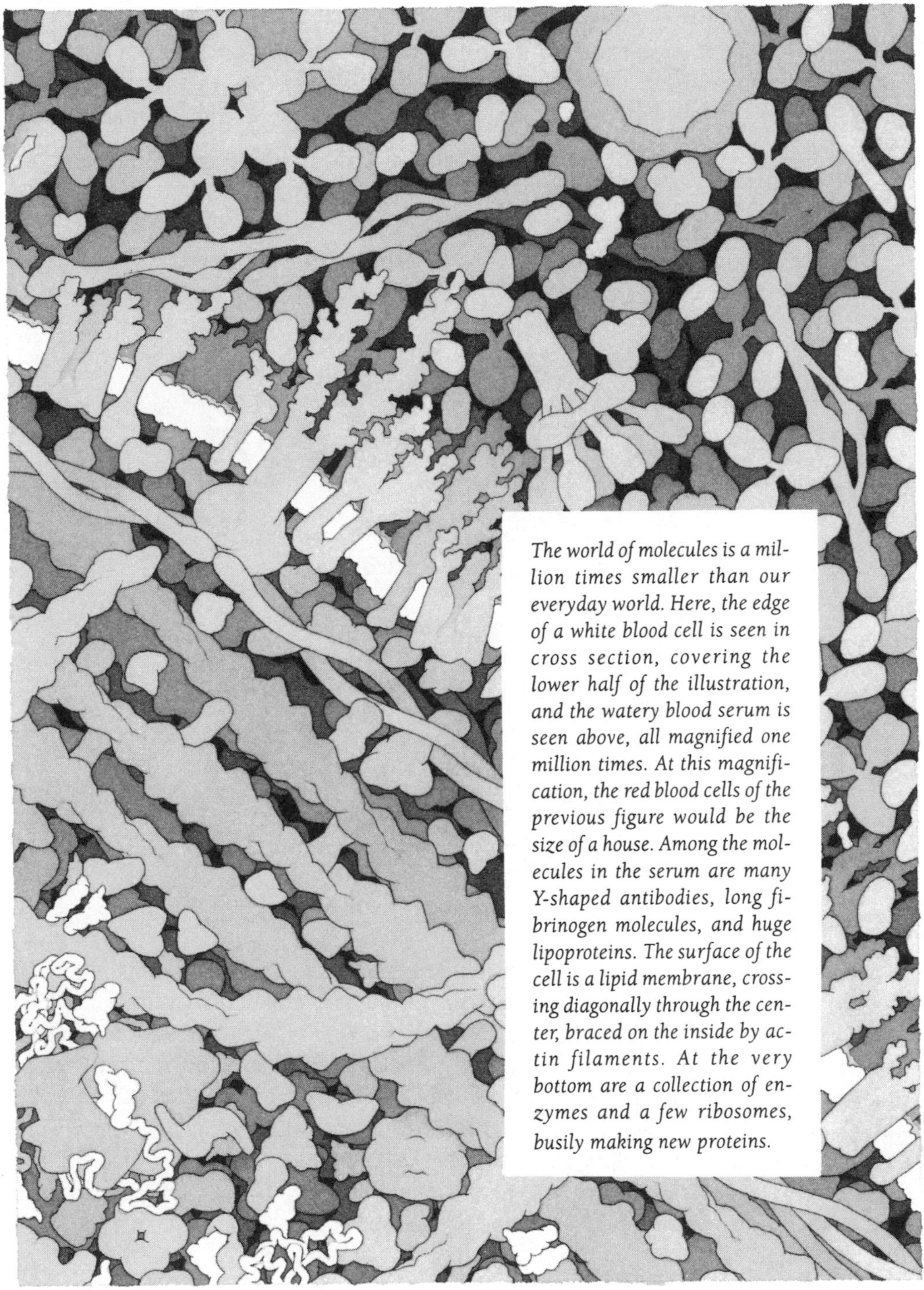

The world of molecules is a million times smaller than our everyday world. Here, the edge of a white blood cell is seen in cross section, covering the lower half of the illustration, and the watery blood serum is seen above, all magnified one million times. At this magnification, the red blood cells of the previous figure would be the size of a house. Among the molecules in the serum are many Y-shaped antibodies, long fibrinogen molecules, and huge lipoproteins. The surface of the cell is a lipid membrane, crossing diagonally through the center, braced on the inside by actin filaments. At the very bottom are a collection of enzymes and a few ribosomes, busily making new proteins.

the specimen on the stage and scatters in a distinctive pattern. The lenses, formed of glass or magnets, then recombine this scattered pattern back into a magnified image of the specimen. Unfortunately, effective lenses for focusing x-rays do not exist, so an x-ray microscope for viewing atoms is not currently feasible. A specimen will scatter a beam of x-rays, but we have no way of recombining them into an enlarged image. This problem has been solved by the technique of x-ray crys-

10,000,000 X   2,000,000 X

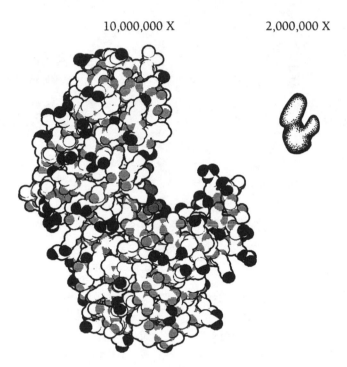

Molecules are illustrated at two consistent magnifications throughout this book. The enzyme phosphoglycerate kinase (page 68) is shown here at both magnifications. Magnified by 2,000,000 times, only the overall shape is apparent, and atoms are too small to be seen individually. Magnified by 10,000,000 times, each atom in this protein is visible as a pea-sized sphere. The coloration of individual atoms is chosen to reflect their chemical nature and the way in which they interact with the surrounding water. Carbon and sulfur atoms are colored white, denoting a weak interaction with water, and nitrogen and oxygen atoms are colored gray, denoting a favorable interaction. Oxygen and nitrogen atoms that carry an electrical charge are colored black, accenting their even stronger interaction with water. For clarity, hydrogen atoms are omitted.

tallography, which is like having an x-ray microscope, except that the lenses are not glass or magnets, but computers. The molecule of interest is placed in an intense beam of x-rays, which are scattered in a distinctive pattern. Since we cannot refocus the scattered rays, the pattern is collected on a sheet of film. Then, in the computer, an image of the molecule is calculated from this pattern. The result is an intimate image of the molecule, revealing the location of every atom. A major limitation of this technique, however, is that a single molecule does not scatter enough x-rays to produce a usable pattern, so a crystal of quadrillions of identical molecules, all in exactly the same orientation, must be used.

Nuclear magnetic resonance (NMR) spectroscopy is also used to create a picture of how each atom is arranged in a molecule. The molecule is probed with radio waves, yielding a distinctive set of signals. Through careful experimental design, the energy of these signals can be interpreted to give a list of atoms that are close neighbors of one another. This list can be used, again with the help of the computer, in an atomic "connect-the-dots" to create an image of the entire molecule.

The illustrations in this book reflect the different levels of detail inherent in each of these techniques. For molecules that have been coaxed into crystalline form or that behave well in NMR experiments, the illustration will show the location of every atom. These molecules are magnified by 10 million times, enlarging each atom to about 3 millimeters across, or the size of a small pea. For other molecules, particularly flexible molecules or molecules with unusual shapes, pictures from the electron microscope provide the best view available, so a simple outline drawing includes all of the information that is currently known. These molecules are drawn five times smaller, at 2 million times magnification, enlarging each atom to slightly larger than a grain of salt.

Many of our own molecules have not been studied at this level of detail. When studying a molecule, researchers typically need large supplies and must choose a source that is rich in the particular molecule. For instance, much of the work on myoglobin has been performed on the protein from sperm whale muscle, because whale muscles contain large quantities of the protein to store oxygen during their extended dives. The pictured examples are representative of our molecules, if not identical to them. Some of the bacterial enzymes pictured here, however, differ significantly from their human counterparts, and are included to illustrate various interesting concepts. Sources for the illustrations are given in the Appendix.

The Names of Molecules

Like animals and plants, molecules are often given both a scientific name and a common or trivial name. The common names, used throughout this book, are often coined by the scientists who first discover and isolate the molecule. They may reflect where the molecule was isolated from, as in asparagine (from asparagus) and guanine (from guano). They may be simple descriptions of the function of the molecule, as in many enzymes: aminotransferase (transfers amino groups), dehydrogenase (removes hydrogen), or synthase (performs one step in a chemical synthesis). Or they may be more fanciful: ubiquitin (found in everything), crystallin (the clear protein in eye lenses), azurin (a brilliant blue protein), or elastin (a rubbery protein).

The scientific names of molecules are detailed descriptions of the way their component atoms are connected, unlike the scientific names of animals and plants, which reflect their evolutionary relationships. For the smaller molecules, a full chemical name may be used, as with 1-palmitoyl-2-oleolyl-phosphatidyl inositol (a fat) or O-β-D-fructofuranosyl-(2→1)-α-D-glucopyranoside (table sugar). From these names, a biochemist can reconstruct the exact position of each atom in the molecule. This type of name becomes impossibly cumbersome, however, for a protein of 1,500 atoms or a DNA molecule composed of millions of atoms. In these cases, a list of the sequence of amino acids in a protein or of nucleotides in DNA (see below) is sufficient to define the chemical structure of the molecule, amounting essentially to a full chemical name. But when we start looking at molecules with carbohydrates attached in branching chains, modified amino acids or nucleotides, and other nonstandard structures, nomenclature can become all but incomprehensible.

Molecular Building Blocks

We require tens of thousands of different molecules to perform the tasks of living. We must be able to build each one exactly when and where it is needed, using only the materials available in the diet. To economize and streamline construction, we build most of these molecules using just a few simple building blocks, arranged in long chains. Just as William Shakespeare, Virginia Woolf, and Emily Dickinson each used twenty-six letters to fashion their distinctive literary works, we build thousands of different molecules, each with a different function, by arranging these building blocks in various orders into chains of different lengths. Simple sugars like glucose are linked to form branching chains of *complex carbohydrate*, often used to protect the outer surfaces of our cells. Four different nucleotides are linked to form sinuous chains of *nucleic acid*, some a hundred million nucleotides in length, in which our genetic information is stored and communicated. Twenty different amino acids are linked to form *proteins*, our molecular machines.

Proteins

Proteins are our most versatile molecules. Long, rigid *structural proteins* form girders and rails, supporting and shaping the body; *motor proteins* move these girders and rails, shipping goods or contracting entire muscles. Tiny *hormones*, composed of only a few hundred atoms, are used as molecular messages, posted from one part of the body to another. *Transport proteins* carry nutrients to hungry cells and cart wastes away. *Enzymes* are the most remarkable proteins; sewn into their chains are a few reactive atoms that perform chemical reactions.

Twenty different amino acids, each with a distinct chemical nature, are used to build our 60,000 different proteins. Some amino acids are small and flexible, some are large and rigid. Some are very soluble in water, others cluster away from water. Some amino acids are chemically reactive, others are nearly inert. By stringing together different combinations of these 20 amino acids, entirely different proteins may be designed. After a long protein chain is built, the surrounding water causes it to fold into a compact shape. Water is a liquid with unusual properties. Water interacts strongly with itself and, like a tight, unfriendly clique, shuns molecules that cannot form as strong an interaction. The portions of a protein that are rich in carbon have little to offer to a water molecule, so they tend to be pushed together, clustering inside when a protein folds. For the same reason, carbon-rich molecules of grease are pushed together, forming droplets on the surface of a watery bowl of soup. Portions of a protein chain that are rich in oxygen and nitrogen, however, interact favorably with water, and tend to stay on the outside. If the protein is designed well, as are all of our component proteins, this wash of conflicting forces balances perfectly to form a unique structure.

This folding, however, is not permanent. Our proteins are delicate machines, easily destroyed by minor environmental insults. Heat can destroy them: when the proteins in an egg are boiled or fried, they unfold and form a stiff, white

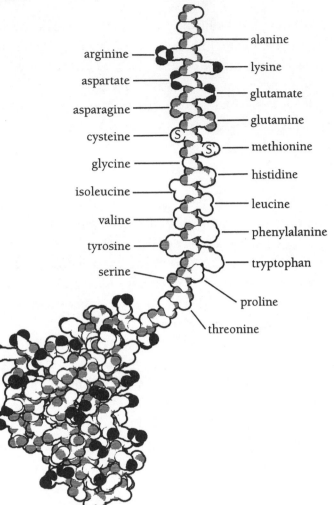

alanine
arginine
lysine
aspartate
glutamate
asparagine
glutamine
cysteine
methionine
glycine
histidine
isoleucine
leucine
valine
phenylalanine
tyrosine
tryptophan
serine
proline
threonine

Proteins are composed of a long chain of amino acids folded into a compact ball. The end of this protein has been teased out to reveal the individual amino acids. The terminal atoms of arginine, lysine, glutamate, and aspartate, colored black, are strongly charged. The lower nine amino acids, on the other hand, are rich in carbon, colored white. The sulfur atoms in cysteine and methionine are marked with an 'S,' and uncharged atoms of oxygen and nitrogen, such as those at the ends of glutamine and asparagine, are colored gray. (10,000,000X) solid. Acid can destroy them: if vinegar and milk are mixed, the proteins in milk will unfold and aggregate, curdling the mixture. Concentrated salt solutions can destroy them: meat and fish are often heavily salted when stored for extended periods without refrigeration, to keep bacterial proteins from digesting them. We create an ideal environment for our proteins, with precisely the right temperature, acidity, and salinity for peak performance.

Alanine is the most abundant amino acid in our proteins, and perhaps the most versatile. It is very small, allowing it to fit into the odd corner. But at the same time, its single methyl group is large enough to lend some stiffness to a chain, shepherding the protein into the typical range of shapes. Alanine is chemically inconspicuous and thus easily accommodated both on the inside of proteins, with the other carbon-rich amino acids, and on the outside of proteins, immersed in water. Alanine is named for acet*al*dehyde, from which it was first synthesized in the laboratory.

Arginine and *lysine* carry long groups with a positive charge at the end. The flexibility of these side chains, and their strong positive charges, make arginine and lysine ideal for recognizing molecules with many negative charges. For example, arginine and lysine extend from the surface of DNA-binding proteins (page 45), attaching them firmly to the phosphates on a DNA double helix (page 36), which each carry a negative charge. The strong positive charge also ensures that both lysine and arginine are found primarily on the surface of proteins, in intimate contact with surrounding water molecules. When they are found in the carbon-rich interior of proteins, as is occasionally necessary for a special chemical function, arginine and lysine will most often be paired with aspartate or glutamate, amino acids that carry a negative charge. Arginine forms brilliant crystals in pure form. Its name comes from the Greek word for "silver," which also yielded the chemical symbol Ag for silver. Lysine was first isolated from artificially digested proteins, and gained its name from the Greek word for "loosening."

Aspartate and *glutamate* each carry an acidic group with a strong negative charge. Glutamate is the longer of the two, with one extra carbon atom, making it slightly more flexible. Both are plentiful in proteins, second only to alanine. They commonly decorate the surfaces of proteins, making them more soluble in water, and are used in specific roles when a positive charge must be recognized. For instance, aspartate and glutamate in calmodulin (page 153) firmly hold four calcium ions, which carry complementary positive charges. Individual glutamate molecules also play an important role as neurotransmitters (page 161), carrying messages in the brain. *Asparagine* and *glutamine* are neutralized versions of aspartate and glutamate, formed by replacing one of the acidic oxygen atoms at the end by an amine (a nitrogen atom carrying two hydrogen atoms). Glutamine, when not used for the fabrication of proteins, is used as the currency of ammonia. Ammonia is an unpleasant, toxic chemical—think of the weak ammonia solution used to clean windows—that cannot be safely circulated through the body in free form. So it is attached to glutamate, forming harmless glutamine, and liberated only when and where it is needed. These amino acids are named according to where they were first isolated: asparagine (and thus aspartate) from asparagus, and glutamate (and thus glutamine) from wheat gluten.

Cysteine and *methionine* each contain an atom of sulfur. Sulfur has several useful roles in cells. The sulfur in cysteine is often used to trap heavy metal ions, such as toxic heavy metals sequestered in metallothionein (page 123). Methionine has a methyl group (a carbon atom carrying three hydrogen atoms) attached to its sulfur atom. In an activated form, this methyl group is used in many reactions in which a new carbon atom is added to another molecule. The sulfur in cysteine and methionine also has several socially unpleasant side effects. Eggs are rich in these amino acids, to provide the growing chick with an abundant supply of sulfur. They may, however, discolor silverware with a tarnish of silver sulfide, or they may cause production of foul-smelling hydrogen sulfide gas during digestion. The name cysteine is derived from the Greek word for "bladder," as it was first isolated from bladder stones; the name methionine is derived from its full chemical name, γ-methylthiol-α-aminobutyric acid.

Cysteine has the unique ability to create a crosslink in a protein. A protein is normally composed of one long chain, folded and packed upon itself. However, if two cysteines are near one another, their sulfur atoms may be chemically connected together, forming internal links in the protein. This glues the protein into the proper shape, helping it resist unfolding. Proteins that act outside of cells, often under unfriendly conditions of acid or salt, typically contain several of these cysteine crosslinks. These include digestive enzymes, such as pepsin (page 59) and chymotrypsin (page 21), and structural proteins, such as keratin (page 92) in hair. Very small proteins, such as the hormone insulin (page 147) also often contain several crosslinks. They are so small that they cannot hold their shape without them.

Glycine is the smallest amino acid, with no side chain at all, making it also the most flexible amino acid. These properties endow it with two structural roles. Glycine fits easily into tight spots, as in the tight triple helix of collagen (page 95). Glycine also forms unusually sharp turns, which are needed in a few places in nearly every protein. One might expect that such a simple and economical amino acid would be used extensively in proteins. This is not the case, however. Too much glycine makes a protein chain overly flexible, so alanine and the larger amino acids typically make up the bulk of proteins. Free glycine has a sweet taste, as noted in its name.

Histidine is the only amino acid that is sensitive to small changes of acidity at the normal neutral pH of the body. Under slightly acidic conditions, it carries a positive charge; under slightly basic conditions, it is neutral. This property makes histidine a powerful chemical tool. Small changes in the placement of histidine within a protein can effect this change, from neutral to charged. Many enzymes have the ability to make this change at will, and use it to force changes in other molecules. In this specialized role, however, only a few histidines are needed in any particular protein, making it the scarcest of the 20 natural amino acids. The name histidine is derived from the Greek word for "tissue."

Isoleucine, *leucine*, and *valine* have branched side chains composed entirely of carbon and hydrogen, making them relatively insoluble in water. These rigid amino acids, along with the three large amino acids described below, are important for the folding of proteins: protein chains fold to shield these amino acids from water, sequestering them inside and leaving the charged amino acids on the surface to deal with the surrounding water. The name leucine is derived from the Greek word for "white," after the brilliant white crystals of pure leucine; the prefix "iso" is derived from the Greek word for "equal," reflecting the chemical similarity of leucine and isoleucine. Valine is a shortened form of "aminovaleric acid."

Phenylalanine, *tyrosine*, and *tryptophan* each contain an "aromatic" group at their ends: a closed, flat ring of bonded atoms. (The term aromatic is derived from the pervasive odor of benzene, a similar molecule composed of six carbon atoms in a ring.) These are the bulkiest amino acids, with their large, inflexible rings. Like the smaller, carbon-rich amino acids isoleucine, leucine, and valine, these amino acids are not overly soluble in water and therefore tend to cluster into the interior of proteins. Phenylalanine carries a six-sided ring similar to that of benzene—an alanine plus a phenyl group. Tyrosine is a phenylalanine with an oxygen atom attached at the far end, named after the Greek word for "cheese," from which it was originally isolated. Apart from its role in proteins, tyrosine is used in the formation of melanin (page 104), the major colorant of skin, hair, and eyes. Several important hormones are also built from tyrosine, including epinephrine, also known as adrenaline, and the thyroid hormones (page 148) used to regulate our overall rate of metabolism. Tryptophan carries a five-sided ring, four carbon atoms and one nitrogen, fused to a six-sided ring of carbon atoms. The neurotransmitter serotonin (page 161) is built from tryptophan. It was first isolated from digests of milk proteins; its name is a combination of the Greek words for "to be broken" and "to appear."

Proline is an unusual amino acid that reaches back and attaches to the protein chain in a second place, forming a closed ring of five atoms. This ring forms a useful kink in the protein chain. Proline is often used to redirect the protein chain back inwards or around a tight corner. Hydroxyproline, a derivatized form of proline, is an important component of collagen (page 95), the major protein in cartilage and connective tissues. Collagen is built with normal proline, then the extra oxygen atoms are added to the mature protein. The name proline is a shortened form of the longer German chemical name.

Serine and *threonine* have a hydroxyl group (an oxygen atom carrying a hydrogen) at their ends. The hydrogen atom on this group is, with a little effort, removable, and is often used in chemical reactions, as in the serine proteinases (page 56). The hydroxyl also makes these amino acids quite soluble in water, so serine and threonine often decorate the outer surfaces of proteins. Serine was first isolated from silk; its name is derived from the Greek word for "silken." Threonine obtained its name from its chemical similarity to the simple sugar threose.

In humans, 8 of these 20 amino acids are "essential" amino acids: tryptophan, methionine, valine, threonine, phenylalanine, leucine, isoleucine, and lysine. All 20 amino acids are absolutely necessary to build new proteins, but, unlike the other 12 amino acids, we cannot make the 8 essential amino acids from scratch. Instead, they must be obtained prefabricated in our diet. Meat, eggs, and cheese are rich sources of all 20 amino acids. But if one is restricted to a vegetable diet, by choice or by circumstance, a mixture of foods is necessary. For instance, corn does not contain appreciable amounts of lysine, so an exclusive diet of corn would lead to severe malnutrition. Beans, on the other hand, are rich in lysine but poor in methionine. A mixture of the two is necessary to provide the entire set of essential amino acids. Traditional combinations have been discovered over the years: corn and beans in the Americas, rice and soybeans in the East.

Nucleic Acids

DNA (deoxyribonucleic acid) and RNA (ribonucleic acid) are not nearly as versatile as proteins, but together, DNA and RNA play a leading role in our cells. Proteins are the workhorses of the cell: they guide each chemical reaction, they provide the structure to hold the body together and the strength to move it. Nucleic acids do not play such a physical role. Instead, nucleic acids carry our library of information. Locked in strands of DNA is the information needed to build each of our proteins: enzymes, molecular motors, protein hormones, and structural proteins.

Nucleic acids are composed of long chains of *nucleotides*. A nucleotide is composed of three parts: a phosphate, a sugar, and a base. To make a DNA or RNA chain, the phosphate of one nucleotide is attached to the sugar of the next, forming an alternating chain: –sugar–phosphate–sugar–phosphate–etc. This sugar–phosphate backbone arranges the attached bases one on top of the next, displaying them in a long stacked row, creating a molecular ticker tape. Four bases are found in DNA, each named after the source from which it was first isolated: *adenine* from pancreas (named after the Greek word for "gland"), *thymine* from thymus, *cytosine* from cells, and *guanine* from bird guano. RNA substitutes the smaller *uracil* base for each thymine and contains one extra oxygen on each sugar. In databases of genetic information, like that of the human genome project, the bases are abbreviated A, T, C, and G—the four letters of the genetic alphabet.

Modified versions of these natural nucleo-

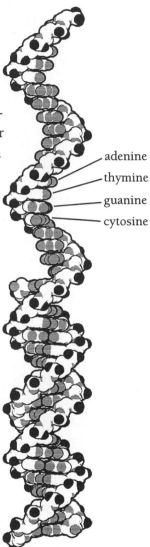

adenine
thymine
guanine
cytosine

Nucleic acids are long, unbranched chains of four different nucleotides, indicated at top. (10,000,000X)

15

tides are effective drugs for stopping the growth of viruses. These drugs contain bases similar to those in DNA, but they have the sugar changed in a clever way. In normal nucleotides, the sugars have two sites that bond to phosphate, allowing them to form the alternating –phosphate–sugar–phosphate–sugar– DNA chain. The molecules used as drugs, however, contain only a single site for attachment, so they act as "chain terminators." When added to a growing DNA chain, they stop further growth because there is nothing onto which to attach the next nucleotide. *Acyclovir*, a modified form of guanine, is used in the treatment of herpes infections. *AZT* (azidothymidine), which carries a thymine base, and *DDI* (dideoxyinosine), which contains the guanine-like base inosine, block the DNA-building machinery of human immunodeficiency virus (HIV), making them useful in the treatment of AIDS.

Carbohydrates

Carbohydrates are familiar in two different forms: simple sugars and complex carbohydrates, or polysaccharides. Simple sugars, such as glucose and fructose, are small and dissolve freely in water. They are easily digestible, providing instant energy, and typically taste sweet. Simple sugars are commonly found in fruits, nectars, and honey, and in purified form in the kitchen. Complex carbohydrates, on the other hand, are composed of dozens to millions of sugars attached in straight or branching chains. Starch and cellulose are two common complex carbohydrates, both of which are composed of long chains of glucose. Complex carbohydrates are often insoluble in water and have little taste. They are used to store sugars, like the starch in corn kernels, or are woven into huge sheets to build structures, like the cellulose in wood.

Glucose is our primary source of power. Glucose is stored in the liver in long, branching chains of *glycogen*, and is released when needed into the bloodstream for delivery to the muscles and brain. The minute-by-minute amounts of glucose delivered in the blood are regulated by the opposed actions of the hormones insulin and glucagon (page 147). Glucose is perhaps the most versatile of sugars. Apart from its ubiquitous role as a power source, it is widely used in a structural role. Wood is given its strength by cellulose, composed of long chains of glucose. Cellulose is completely indigestible by humans: only a few specialized microorganisms have been successful in releasing the sugars from wood. A similar chain of N-acetylglucosamine, a modified form of glucose, forms the tough *chitin* coat of insects.

Fructose is sweeter than glucose and is widely used as a commercial sweetener. Surprisingly, in spite of its central role in the powering of cells, glucose tastes a bit too flat for most people's preferences. Table sugar, with its more palatable sweet taste, is *sucrose*, a disaccharide composed of glucose and fructose. Fructose is common in fruit, giving it its name, and is often the major sugar in honey, since plants entice bees to their flowers with a concentrated solution of fructose nectar. Fructose is made in substantial quantities in only one place in the human body. It is made by the cells surrounding the seminal vesicles, and is used as the major source of energy for swimming sperm. Since the cells surrounding the seminal vesicles and the cells of the oviduct do not themselves use fructose to any great extent, the fructose remains available exclusively for use by the sperm.

Galactose is most commonly found in *lactose*, the sugar found in milk. Lactose is a disaccharide, like sucrose, composed of galactose and glucose. During digestion, lactose is broken in half and the pieces are used for energy. Lactose-intolerant individuals cannot perform this cleavage, and suffer the consequences of undigested lactose (see page 62). Galactose is also the molecule responsible for the ABO blood group. The surfaces of blood cells are decorated with short chains of sugars. In some individuals, the last sugar in the chains is galactose, in other people it is N-acetylgalactosamine (a modified form of galactose). These correspond to types B and A, respectively. People with type AB blood have some of each, and those with type O blood have neither. Proper matching of blood type is critically important in blood transfusions. If type B blood is transfused into a patient with type A blood, the patient's immune system will identify the galactose on the type B blood as being foreign, and will destroy it. Thus, type B blood cannot be given to type A patients, and vice versa. Type O blood does not carry either sugar, so it may be given to type A, type B, and type AB patients. Unfortunately, type O patients

Polysaccharides are long, often-branched chains of simple sugars, such as glucose or galactose. (10,000,000X)

cannot take type A, type B, or type AB blood—they see all as being foreign. Thus, people with type O blood are universal donors, but can only accept transfusions from other type O donors.

2
Building Molecules

The secret of life, discovered in terrestrial seas 4 billion years ago, is the ability to build molecules according to need. This ability separates the animate world from the inanimate; it quickens us with life. When one of our proteins becomes damaged, we build a replacement; if a cell membrane is damaged, we make repairs. If natural resources are plentiful, we make extra molecules and grow, building muscle and gaining the competitive advantages of larger size and greater strength. When a new food becomes available, we make the specialized molecules needed to digest it. When bacteria attack, our blood cells make toxic molecules to kill them. And ultimately, a single cell, when paired with an appropriate mate, can build an entirely new human being, molecule by molecule, identical in every basic way to ourselves.

The vital discovery made at the dawn of life was the method of storing molecular information: the information needed to construct all of these different molecules. Coded in strands of DNA are exact instructions for building each of our thousands of *proteins*. This is valuable information, inherited directly from our parents. It is used every second of life, since proteins perform nearly all of our bodily functions. Proteins support and connect our cells, giving us form and shape; proteins mediate communication throughout the body, and between the body and the environment; and proteins perform all of the chemical reactions needed for living.

19

All of our other molecules are then fabricated by these proteins. Enzymes, proteins that catalyze chemical reactions, build the carbohydrates, the lipids, the nucleic acids, and all of our other nonprotein accessories, using the raw materials available in the diet. In this chapter, we explore first the diversity of our personal enzymes, and then the machinery that builds them and other proteins according to the plans stored in our DNA.

Enzymes

Enzymes are molecular machines: machines made of protein that perform an atomic task, efficiently and accurately. Thousands of different enzymes are at work in each of us at this moment. Some are connecting molecules together, building amino acids or nucleotides from their component parts. Some are breaking molecules in half, digesting long protein and carbohydrate chains into usable pieces. Some enzymes are plucking individual atoms from one molecule and placing them on another, adding oxygen to make a molecule more soluble in water, or carbon to make it less so. Others are shuffling atoms around inside a single molecule, changing one sugar into another or transforming a poison into a harmless derivative.

Enzymes are large but inconspicuous molecules. They are typically globular in shape, most often with a cleft on one side. A few key amino acids, precisely arranged inside this cleft, perform the chemical task. Three glutamates, each carrying a negative charge, may be arranged in a triangle to snare a magnesium ion, which carries a complementary positive charge. Large, carbon-rich phenylalanines and tryptophans may line a deep pocket, forming a comfortable nest for a fatty tail; a tyrosine may form a door at its surface, trapping the target molecule inside while it is broken for use as food. An array of arginines may immobilize an acidic molecule, forcing a susceptible atom against a reactive serine. This cleft, with its perfectly tailored arrangement of amino acids, is termed the *active site*.

The novelty of these enzymes, developed by living cells to solve chemical problems, cannot help but surprise and astound. The majesty of nature is revealed in perhaps two places in our world: in the diversity of species inhabiting the Earth, evolving to fill every ocean and every piece of land, and in the enzymes of molecular fabrication, ceaselessly working within each of us. With a variegated collection of enzymes, each formed of 20 simple amino acids and a few dozen helper molecules, we turn food into flesh and water into blood.

Chymotrypsin

Chymotrypsin is arguably the best understood of all enzymes. It is a small digestive enzyme, with about 245 amino acids, that is secreted by the pancreas to aid in the digestion of proteins. Chymotrypsin was one of the first enzymes to be studied by x-ray crystallography, so the exact position of its every atom is known. Each chemical step that it performs has been studied in intimate detail, and the exact amino acids orchestrating each step have been located and explored. The picture that emerges is remarkable. Chymotrypsin is a molecular machine with no moving parts, but its simple cleavage reaction—breaking protein chains at points next to their large amino acids—requires the exact placement of dozens of amino acids. The entire active site cleft is lined with precisely arranged amino acids, which stretch out and immobilize a susceptible chain in the protein victim. At the center of the active site is the "catalytic triad": the three key amino acids that perform the cleavage reaction. A serine attacks the target protein bond, and a histidine and aspartate together activate it. An "oxyanion hole," composed of two nitrogen atoms perfectly placed adjacent to the catalytic triad, helps to stabilize the chemical intermediate formed in the cleavage, smoothly guiding the reaction from start to finish. A "specificity pocket," a deep hole lined with carbon-rich amino acids, makes sure that only regions of a protein with large amino acids are recognized and cleaved.

Looking at chymotrypsin, and at the even larger enzymes on the following pages, one might ask why enzymes are so large, given that only a few dozen

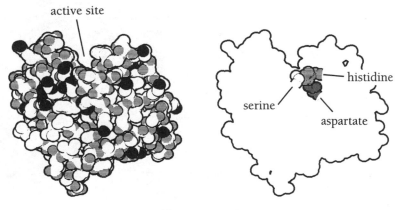

Chymotrypsin

amino acids are necessary to perform the chemical reaction. This enormous overhead of structure is needed to arrange these key amino acids in exactly the right relative orientation. The problem—successfully solved by each enzyme—is to start with a long chain of protein, and to have it fold into a rigid, defined shape, with these few amino acids arranged in the proper orientation every time. The smallest enzymes seem to require about 150 amino acids to perform this feat. Many enzymes are much larger than this, with hundreds to thousands of amino acids.

Carbonic Anhydrase

Carbonic anhydrase is a perfect enzyme: it performs its chemical reaction as fast as is physically possible. As with most enzymes, the chemical reaction catalyzed by carbonic anhydrase occurs naturally by itself, but only very slowly. The enzyme combines carbon dioxide and water to form bicarbonate ions, speeding up the natural reaction over 10 million times. One carbonic anhydrase enzyme can dissolve hundreds of thousands of carbon dioxide molecules every second. The speed of its reaction is limited only by the speed with which carbon dioxide can bump through the surrounding water into the active site of the enzyme.

The power of enzymes lies in the precise control they exert. With carbonic anhydrase, we convert carbon dioxide into soluble bicarbonate at will, precisely when and where it is needed. Carbon dioxide is formed in large quantities during respiration. To keep it from bubbling out of the blood like the carbonation in soda, carbonic anhydrase combines it with water to form bicarbonate ions, which are freely soluble in water. In the lungs, the process is reversed, again with the help of carbonic anhydrase, releasing carbon dioxide as we exhale.

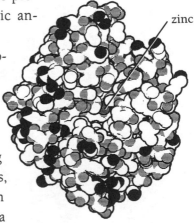

zinc

A zinc ion, held tightly in the active site, provides the extra chemical leverage needed for the carbonic anhydrase reaction. Metal ions play many roles in the body. Magnesium and calcium, which carry a stable double positive charge, are often used to hold molecules with negative charges, positioning them in the enzyme's active site. Zinc and iron atoms, with their stronger charges, are often used to weaken susceptible bonds in target molecules. Iron also has a strong affinity for oxygen gas, and is used in large

Carbonic Anhydrase

quantities in the blood as a carrier of oxygen (see hemoglobin, page 75). Iron and copper, and unusual metals such as molybdenum and vanadium, easily pick up and release electrons, changing charge. In proteins, they are often held inside cages of sulfur and used to shuttle single electrons from one molecule to another.

Aspartate Carbamoyltransferase

Simple enzymes like carbonic anhydrase and chymotrypsin are always active. Whenever carbon dioxide bumps into carbonic anhydrase, the enzyme performs its chemical change. Chymotrypsin clips *any* protein chain it touches. But, inside our cells, not all enzymes can act similarly. If every enzyme was active at all times, a mad rush of chemical change would result. Enzymes that build nucleotides or amino acids would be active at the same time as those that digest them, resulting in a futile cycle of synthesis and destruction. To avoid this problem, key enzymes are carefully regulated. They are activated only when their products are needed, and shut down otherwise.

Dozens of enzymes are needed to make the DNA bases cytosine and thymine from their component atoms. The first step is a "condensation" reaction, connecting two short molecules to form one longer chain, performed by aspartate carbamoyltransferase. Other enzymes then connect the ends of this chain to form the six-sided ring of nucleotide bases, and half a dozen others shuffle atoms around to form each of the bases. In bacteria, the first enzyme in the sequence, aspartate carbamoyltransferase, controls the entire pathway. (In human cells, the regulation is more complex, involving the interaction of several of the enzymes in the pathway.) Bacterial aspartate carbamoyltransferase determines when thymine and cytosine will be made, through a battle of opposing forces. It is an *allosteric* enzyme, referring to its remarkable changes in shape (the term is derived from the Greek for "other shape"). The enzyme is composed of six large catalytic subunits and six smaller regulatory subunits. The active site of the enzyme is located where two individual catalytic subunits touch, so the position of the two subunits relative to one another is critical. If the two subunits are in tight contact, an amino acid from one extends into the active site of the other, blocking its action. If the two are pulled slightly apart, however, the active sites are revealed, allowing molecules to enter and the reaction to be performed. This is the job of the regulatory subunits: they alternately pull the central catalytic subunits apart, turning the enzyme on, or allow them to stick together, turning the entire complex off.

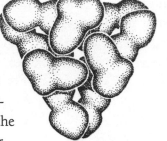

This is a powerful capability. When the proper raw materials are in plentiful supply, they bind to the active sites and force the molecule into the open, active shape. Notice that if a single molecule binds to one catalytic subunit, it forces open the entire interconnected complex, activating the other five subunits at the same time. The regulatory subunits, however, bind the nucleotide CTP (cytidine triphosphate), the final product of the entire collection of enzymes. If CTP is plentiful, it binds to a regulatory domain, triggering it to close the enzyme down, shutting each active site off. Again, binding at one or two sites will shut down the entire enzyme, closing all six sites. What results is a tug-of-war between supply and demand. When raw materials are more plentiful, the enzyme is turned on; when there is enough of the ultimate product, the enzyme is turned off.

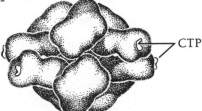

CTP

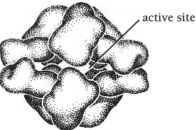

active site

Take just a moment to ponder the immensity of this enzyme. The entire complex is composed of over 40,000 atoms, each of which plays a vital role. The handful of atoms that actually perform the chemical re-

Aspartate Carbamoyltransferase

action are the central players. But they are not the only important atoms within the enzyme—every atom plays a supporting part. The atoms lining the surfaces between subunits are chosen to complement one another exactly, to orchestrate the shifting regulatory motions. The atoms covering the surface are carefully picked to interact optimally with water, ensuring that the enzyme doesn't form a pasty aggregate, but remains an individual, floating factory. And the thousands of interior atoms are chosen to fit like a jigsaw puzzle, interlocking into a sturdy framework. Aspartate carbamoyltransferase is fully as complex as any fine automobile in our familiar world. And, just as manufacturers invest a great deal

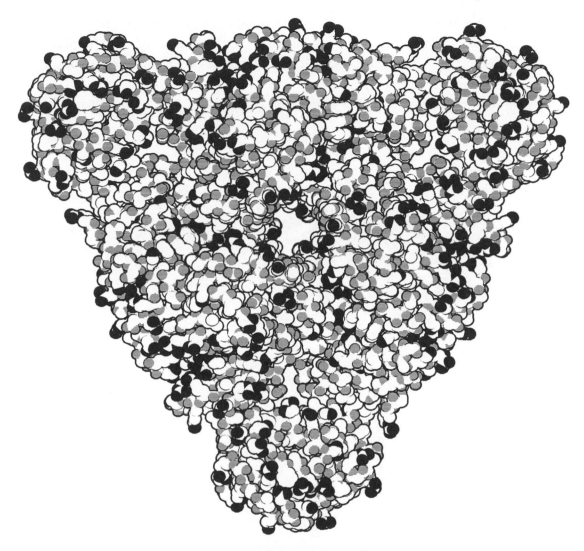

Aspartate Carbamoyltransferase

of research and time into the design of an automobile, enzymes like aspartate carbamoyltransferase have been finely tuned over the course of evolution.

Tryptophan Synthase

Tryptophan synthase is one of the wonders of the molecular world. It is a complex of two enzymes that perform the last two steps in the synthesis of tryptophan. The first step is performed by two copies of the smaller enzyme, located at the far ends of the complex. They build *indole*, the side chain of tryptophan. The second step is performed by two copies of the larger enzyme located at the center. It removes the hydroxyl group from a serine and adds the indole in its place, forming tryptophan. (The tryptophan synthetase shown here is a bacterial

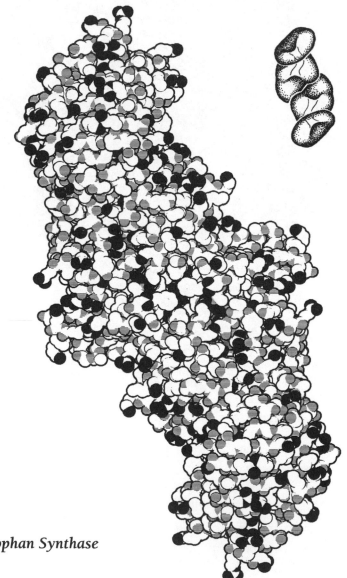

Tryptophan Synthase

enzyme. Tryptophan is an essential amino acid for human beings—we cannot make it from scratch, because we do not make this enzyme.)

The wonder of tryptophan synthetase lies in the interaction between these two steps. One might imagine that the indole formed in the first step, at the ends of the enzyme, would be released and would have to find its way to the center to complete the reaction. Of course, some might be lost, left to scatter randomly throughout the cell. But tryptophan synthase is much cleverer than this. Inside the enzyme, there is a narrow tunnel connecting the outer enzymes with the inner, just wide enough for indole to push through. Indole is delivered directly from the first enzyme to the second, with no chance for escape. This *substrate channeling* speeds up the entire reaction by a thousandfold.

Many of our own enzyme pathways utilize this efficient technique. The enzymes that build cytosine and thymine bases, including aspartate carbamoyltransferase, form a large complex, passing intermediate forms of the nucleotides from active site to active site. Two enzymes of the citric acid cycle (page 70) are actually complexes of enzymes, designed to perform three reactions in rapid succession. Many researchers believe that more tenuous connections may also be important. For instance, the 10 glycolytic enzymes (page 68) are thought by some to form a "glycosome," clustering together into a molecular assembly line.

Aspartate Aminotransferase

Many enzymes use tools to assist in their chemical reactions. These tools may be metal atoms, such as the zinc atom used by carbonic anhydrase, or they may be small, reactive organic molecules. Occasionally, these tools are quite unusual and are difficult for us to fabricate. In these cases, starting materials for building the tools must be obtained in the diet as *vitamins*. The B-complex vitamins, in particular, are used as enzymatic tools. Vitamin B_1 (thiamin) helps enzymes coax carbon dioxide away from acidic molecules. Vitamin B_2 (riboflavin) and niacin are used to build carriers of hydrogen atoms, important in the conversion of food into usable chemical energy. Vitamin B_6 (pyridoxine) is used by enzymes that shuffle nitrogen atoms between molecules.

Aspartate aminotransferase is an enzyme that relies on the chemical leverage of vitamin B_6. An *aminotransferase*, as the name implies, moves amines from one molecule to another—aspartate aminotransferase shuffles amines be-

27

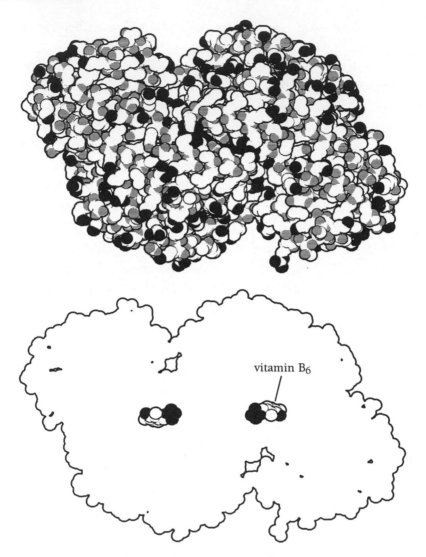

Aspartate Aminotransferase

tween different amino acids. The reaction is a double displacement reaction, often referred to as a "ping-pong" reaction. Vitamin B_6 is held within the active site. When an amino acid, such as aspartate, binds next to it, a reactive aldehyde group on the vitamin extracts its amine—Ping!—and the clipped molecule is released. A second molecule then enters the active site, and the amine is attached to it—Pong! When not actively performing a reaction, the enzyme protects its valuable tool by attaching it to a nearby lysine, storing it out of harm's reach.

Dihydrofolate Reductase

Many enzymes were developed very early in the evolution of life and have not been substantially improved in billions of years and trillions of generations. These enzymes are very similar in all organisms, from bacteria to plants to human beings. Small differences may exist, caused by random mutations from generation to generation. A few amino acids may be changed or a small loop may be added, but the functional machinery remains the same. The enzyme dihydrofolate reductase is a good example. It builds *tetrahydrofolate*, a small molecule needed for moving carbon atoms. Our enzyme contains 186 amino acids, and that from bacteria is similar, but smaller, with about 160 amino acids. Both fold into the same shape, with a large active site cleft on one side. They differ only in a few decorations on their surfaces.

The antibiotic drug *trimethoprim* acts by blocking dihydrofolate reductase, killing bacteria by depriving them of tetrahydrofolate. One might think that this is not a logical target for a drug: both humans and bacteria contain the enzyme, so the drug might kill the patient along with the disease. The basis for the action of trimethoprim lies in the differences between the bacterial enzyme and the human enzyme. The drug focuses on these differences, ignoring the similarities. Consequently, trimethoprim is about 30,000 times more effective at blocking the bacterial enzyme than the human enzyme. A low dose of the drug kills bacteria, while leaving the patient relatively unharmed.

Of course, an even more effective antibacterial drug would be one that attacked an enzyme *only* present in bacteria. The *sulfa drugs* were the first widely used antibiotics that took advantage of this idea. We make tetrahydrofolate from the vitamin folic acid, which is common in meats and vegetables. But bacteria make their own folic acid, because they rarely find a source of it in their limited diets. Thus, the bacterial enzymes that make folic acid are excellent targets for antibacterial drugs. Folic acid is composed of three

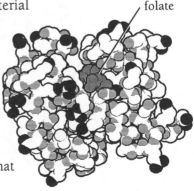

folate

Dihydrofolate Reductase

groups: PABA (*p*-aminobenzoic acid) at the center, with a glutamate at one end and an unusual pteridine ring at the other. (PABA also finds common use in sunblocks.) The sulfa drugs, such as sulfanilamide, are chemically similar to the PABA group at the center of folic acid and bind tightly into the enzyme that builds folic acid from PABA, blocking its action. Since we do not make folic acid from scratch, obtaining it exclusively in the diet, the drug does not poison us. It kills only bacteria, as they try to make their vital supplies.

Thymidylate Synthase

The four RNA nucleotides—adenine, uracil, cytosine, and guanine—are needed in plentiful supply every day, since the minute-to-minute reading and usage of genetic information is performed by RNA (page 42). RNA is continually made and remade as new proteins are needed. New DNA, however, is made at only one time in the life of a cell: just before it divides, forming two daughter cells. Adenine, cytosine, and guanine are also used to make DNA, but the fourth base is different. Uracil is used in RNA, and the similar base thymine is used in DNA. Thymine nucleotides are needed in this single role: as building blocks for DNA. Because thymine is essential for growth, the enzymes involved in its synthesis, such as thymidylate synthase, are attractive targets for anticancer drugs.

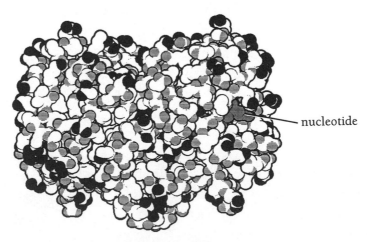

Thymidylate Synthase

Cancer is an insidious disease, formed when normal cells run wild, proliferating without control. Cancer has its genesis in many environmental hazards—ultraviolet radiation, reactive chemicals, and some viruses—that disrupt the normal controls on growth. Chemotherapy walks a fine line, attempting to kill these rogue cells without poisoning the patient in the process. Many chemotherapeutic drugs are poisons that specifically attack *growing* tissues. For instance, a drug that blocks thymidylate synthase will stop production of thymine, so new DNA cannot be constructed, blocking the reproduction of cells. Unfortunately, these drugs also kill any other normally growing cells, leading to the severe side effects of cancer chemotherapy. As the growing cells in hair follicles die, hair falls out. As the cells lining the stomach die, terrible digestive problems ensue. As the cells in the bone marrow die, the blood weakens through lack of new blood cells. The power of chemotherapy is inescapable, however: treatment with enzyme-blocking drugs such as *methotrexate* can lead to the long-term remission of some types of cancer and leukemia.

Glutamine Synthetase

We require a constant supply of nitrogen to build the bases in nucleic acids and the amino acids in proteins. The nitrogen gas in the air and the nitrogen in nitrates and nitrites, although abundant, are not reactive enough for this use. Ammonia is the preferred source of nitrogen for these reactions. Unfortunately, ammonia is very toxic and cannot be stored or transported safely. Instead, ammonia is attached to the amino acid glutamate, forming glutamine. Because it is a natural amino acid, normally used to build proteins, glutamine is easily transported throughout the body in large amounts. Ammonia may then be liberated only when needed.

Glutamine synthetase connects a molecule of ammonia to the amino acid glutamate. A molecule of ATP (adenosine triphosphate) (page 67) is used to power the process, to ensure that the reaction is performed only in the proper direction and not in reverse, carelessly liberating poisonous ammonia. The human enzyme is quite simple, but the bacterial enzyme shown here is a highly regulated allosteric enzyme. Glutamine synthetase has been likened to a molecular computer. With its 12 interacting subunits, arranged in two rings of six, it senses the amounts of the amino acids and nucleotides ultimately constructed from the ammonia in glutamine. Glutamine synthetase weighs the concentrations of each, computes

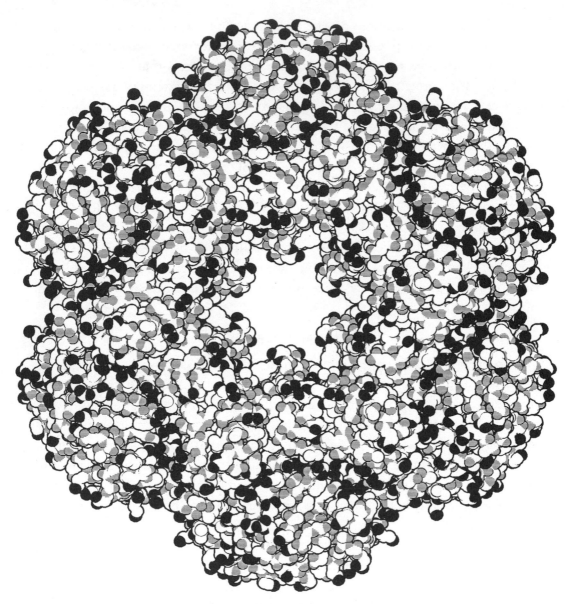

Glutamine Synthetase

whether there is an overall deficit or excess, and turns on or off based on the result.

Nitrogen is found everywhere on Earth, forming about three-fourths of the air. Nitrogen gas, however, is chemically inert and of little use to us. Our primary source of nitrogen is the ammonia in amino acids and nucleotides, obtained by eating other living things. But small amounts of ammonia are lost from the biosphere over time, locked up in minerals and buried out of reach. To replenish the global supply of biological nitrogen, nitrogen gas is converted into ammonia in the process of *nitrogen fixation*. Today, this is accomplished in three ways: about 15% is formed geologically, by lightning and ultraviolet radiation; 25% is produced industrially and distributed as fertilizer; and the remaining 60% is produced by a small class of bacteria and algae. These "diazotrophic" microorganisms fix nitrogen using *nitrogenases*, enzymes that rip apart the two tightly bound atoms in nitrogen gas and add hydrogen atoms to them, forming ammonia. Nitrogenases contain dozens of reactive iron atoms, as well as rarer metals such as molybdenum and vanadium. These unusual metal ions are required to apply the chemical tension that wrenches apart the stable nitrogen molecule. However, they are extremely sensitive to oxygen. Leguminous plants, like peas and beans, have worked out a solution to this problem. In a classic example of symbiotic cooperation, legume roots build a nodule custom-made for bacteria, filled with *leghemoglobin*, a protein similar to the hemoglobin (page 75) that carries oxygen in our blood. Leghemoglobin soaks up any oxygen that ventures near. In return for this safe haven, the bacteria release some of their fixed nitrogen for use by the plant. This abundant supply of ammonia carries a heavy price, however. Nitrogen fixation is very expensive, requiring about 16 ATP molecules per nitrogen molecule split into ammonia.

Alcohol Dehydrogenase

The alcohol in wine, beer, and distilled liquors, although not normally consumed for its calories, is broken down in the liver to yield energy. First, hydrogen atoms are removed from alcohol and transferred to NAD, a small carrier molecule made from the vitamin niacin. This converts the alcohol into acetaldehyde, a highly toxic substance. Acetaldehyde is then converted into acetic acid, the major component of vinegar, through the action of additional enzymes. Both the hydrogen atoms, carried on NAD, and the acetic acid then feed into the normal energy-producing mechanism (see Chapter 2). Normal adults have enough alcohol dehydrogenase in the liver to convert about 10 millileters of alcohol—a typical drink—each hour. Heavy drinkers may build up more of this enzyme, reducing their susceptibility to alcohol. Even the staunchest teetotaler,

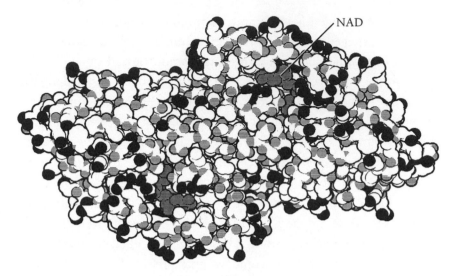

NAD

Alcohol Dehydrogenase

however, needs alcohol dehydrogenase, because alcohol is produced naturally by bacteria in the intestines.

The protection provided by alcohol dehydrogenase is a double-edged sword—it can backfire with deadly results. Alcohol dehydrogenase will also break down other alcohols. It converts methanol, or wood alcohol, into formaldehyde, the substance commonly used for embalming. Formaldehyde is a toxic, reactive molecule that glues portions of proteins together, stopping the delicate motions that are necessary for their action. Small amounts of methanol cause blindness as the formaldehyde formed in the retina paralyzes the delicate motions of the light-sensing opsin proteins (page 154). Larger amounts—a small glassful—cause widespread tissue damage and death.

D-Xylose Isomerase

D-Xylose isomerase plays a key role in the food industry. Isomerases are enzymes that shuffle atoms around in a molecule, as opposed to adding new chemical groups or breaking molecules into pieces. D-Xylose isomerase, using an atom of magnesium, catalyzes the interconversion of different simple sugars. Its industrial utility stems from its ability to convert glucose into fructose. Large quantities of glucose are produced from the breakdown of corn starch by the digestive enzyme

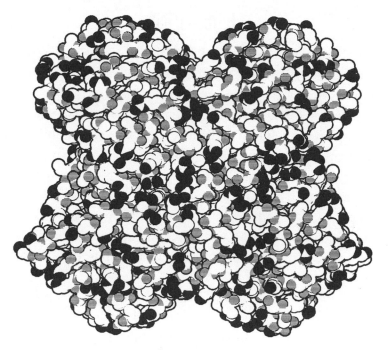

D-Xylose Isomerase

amylase (page 61). Glucose, however, does not have the taste that consumers prefer, so it is converted to sweeter fructose by application of bacterial d-xylose isomerase.

Enzymes are remarkable machines, and like machines of our familiar world, they are often the focus of industrial entrepreneurs. D-Xylose isomerase is but one example. Digestive enzymes have found use in laundry detergents, chewing away nasty stains, and in cleaning solutions for contact lenses. Other digestive enzymes, purified in large quantities from bacteria, curdle milk into cheese. Purified bacterial enzymes are used routinely in biochemical laboratories to rewrite genetic information, engineering new strains of bacteria and developing faster-growing and disease-resistant crops. Many other exciting possibilities are currently being explored. If the nitrogenase enzyme could be harnessed industrially, ammonia fertilizers could be made cheaply and efficiently. Designer enzymes could direct the building of new medicinal drugs. Clever genetic engineering may even give the world, after centuries of searching, the elusive blue rose.

Information

The synthesis of proteins is the primary business of our cells. Nearly every cellular task is performed by proteins; the enzymes shown above are only a few spectacular examples. We routinely construct tens of thousands of different proteins, repair them when necessary, and replace them when worn. But whereas each type of small molecule in the cell—ATP, alanine, glucose, etc.—is built by a different, specialized enzyme, all of our proteins are made by one set of molecular machinery. The key that makes this possible is the ability to build proteins according to a molecular blueprint. To construct a new protein, we do not need to build a new set of enzymes to put it together one piece at a time. Instead, we merely look up the recipe in the genetic blueprint, and build the protein accordingly. For all living organisms on the Earth, from bacteria to human beings, this blueprint is stored in strands of DNA.

Each of our cells contains the information needed to reproduce all of its own component proteins, encoded in a string of nucleotides. DNA carries the hereditary information passed from parent to child, from generation to generation: our personal inheritance, forming a living link to the earliest cells on Earth. The recipe for tyrosinase (page 104)—which makes the melanin that colors eyes brown or blue, hair from blond to glossy black, and skin in shades from deep brown to pale pink—is only the most visible of this hereditary information. Written in strands of DNA are the instructions to build an entire body. Implicit in these strands is the entire history of evolution. The instructions have been copied, edited, shuffled, rewritten, written over, and corrected over billions of years, yielding the densely typed volumes now present in each of our cells.

DNA (Deoxyribonucleic Acid)

DNA is perhaps the most beautiful of our molecules, but like a fine book, its true beauty lies not in its binding, but in the words written within. DNA is composed of two long chains of nucleic acid, wrapped around one another in a double helix. The sequence of bases in each DNA strand—the order of adenines, thymines, cytosines, and guanines—is the genetic text, holding the information needed to build each of our proteins. Human cells contain 46 pieces of DNA, each a few centimeters long and vanishingly narrow. Together, they contain about 6 billion nucleotides. Likening each nucleotide to a letter, each cell contains the information in several thousand books. Each page contains enough information to make a protein.

Parents each provide half of the set of 46 strands, so our cells contain two similar sets of 23 strands. The choice between reading the mother's or the father's version in each particular case provides the rich mixture of inherited traits seen in children. Features that may be traced to a single protein show very specific rules of heredity: if neither parent provides the recipe for tyrosinase (page 104), the child will be albino. Other personal features—shape of the nose and eyes, build of body, temperament—may depend on the combined action of many regions of DNA, leading to subtle mixtures of recognizable traits from each parent.

Surprisingly, less than one-tenth of our DNA carries instructions for proteins. The remainder is filled with a jungle of unusual sequences. *Satellite sequences*, perhaps a dozen nucleotides in length, are repeated hundreds or thousands of times in a row, creating vast stretches of homogeneous DNA. *Transposable sequences*, a few hundred nucleotides in length, carry the ability to build copies of themselves at distant locations. These sequences have multiplied themselves a millionfold and are now colonized throughout our chromosomes. These large, empty regions, combined with these mobile units, may provide the landscape on which new proteins are designed. As transposable elements jump from site to site, they occasionally carry fragments of the nearby genes with them; when they find a new home, these fragments may be combined with another gene, piecing together a new protein.

DNA

Many organisms do not have this luxury. Bacteria are highly competitive, constantly racing to multiply and to consume any new source of nutrients. At peak growth, they may divide every half-hour, and therefore it is advantageous to keep their amount of DNA to a minimum so that it may be rapidly copied with each generation. Thus, bacterial DNA is filled with instructions for proteins, with little wasted space. Viruses may be even more parsimonious. Viral DNA strands (or in some types, RNA) must be very compact in order to fit into their tiny protein shells. In some cases, there is simply not enough room to fit all of the necessary information into their short lengths of DNA. Some of the instructions are actually overlapped. Instructions for the end of one protein also serve as the instructions for the beginning of the next, just as the end of this sentence is the beginning of *the last sentence* contains the end of the previous sentence.

p53 Tumor Suppressor

The exact order of bases in our DNA must be conserved over an entire lifetime. Any changes in the order of bases, termed *mutations*, could result in the building of different proteins, perhaps defective proteins. Simple damage to DNA is repaired by a horde of repair enzymes, which clip out the defective bits and restore the proper sequence of bases. But greater damage calls for more drastic measures. Cells with extensive damage to their DNA are normally destroyed. In response to irreparable damage, cells make the *p53 tumor suppressor*, a protein normally present in only small quantities (p53 refers to the molecular weight of the protein—about 53,000 times that of a single hydrogen atom). The p53 tumor suppressor is composed of four identical protein subunits connected by flexible tethers. Together, they wrap around the DNA double helix and block regions that carry information on the timing of cell division. Damaged cells make large quantities of the protein, which binds to the DNA, blocking the normal signals. The cell cannot divide and, ultimately, dies. Its damaged genetic material is no longer a threat. If this system fails, however, the result is deadly.

Most cancers are the direct result of failure in the p53 tumor suppressor system. Half of the current incidence of cancer is caused by a mutation in the DNA instructions for building this protein. A faulty protein is made, which allows damaged cells to slip through the division cycle. These damaged cells multiply without control, pushing normal, healthy cells out of the way, forming a cancerous growth. Many other cancers are caused by viruses that attack p53 tumor suppressor, with the same effect. They build a protein that blocks the p53 tumor suppressor, leading to uncontrolled growth of the infected cells and cancer. The development of drugs to fight these forms of cancer is particularly difficult. Most other drugs attack a given protein, stopping its action.

p53 Tumor Suppressor

Aspirin attacks cyclooxygenase (page 143), blocking the formation of pain signals. Penicillin blocks the enzyme that builds the tough cell walls of bacteria, leaving them defenseless. But in the case of p53 tumor suppressor, the goal is to find a drug that will restore a faulty enzyme. This is a far more difficult task, which has yet to be solved.

Nucleosome

DNA is many millions of times longer than it is wide, making it susceptible to the constant shearing forces of surrounding molecules. In the nucleus, the delicate strands of DNA are stored and protected in *chromosomes*, so named because they stain colorfully and are easily seen in the microscope. Each of our 46 chromosomes contains a single strand of DNA, compactly packaged by millions of protein molecules. Chromosomes are in their most compact, visible form when a cell divides. The familiar X-shaped chromosome is actually two identical chromosomes, photographed just before they are separated, one chromosome for each of the daughter cells.

    Inside chromosomes, DNA is spooled into *nucleosomes*. A nucleosome contains about 15 turns of the DNA helix—a length of about 150 nucleotides—wrapped twice around a protein core. The core is composed of eight *histone* proteins, which contain a large number of positively charged amino acids, perfectly complementing the negatively charged phosphates on the DNA. When the information written in any particular stretch of DNA is needed to build a protein, the nucleosome unwinds, freeing the DNA for reading. Afterwards, it is coiled back around the histones, where it is stored and protected until the next time it is needed.

Topoisomerase

DNA strands are packed into a space many thousands of times smaller than their length. DNA must be manipulated within this tiny space, causing a number of topological problems. When DNA is unwound for reading, the regions downstream become tightly overwound, and the pressure must be relieved. Think of trying to pull apart the individual strands in a length of rope—it quickly becomes snarled. A larger problem occurs when the cell divides. Just before division, the nucleus con-

Nucleosome

39

tains two copies of each of the 46 strands of DNA. Each pair of strands must be untangled, to allow them to separate, one into each of the daughter cells.

Topoisomerases solve these topological nightmares. The stresses of overwinding are easily relieved. A topoisomerase breaks one strand of the DNA helix, allowing the DNA to unwind and relax. The strand is then reconnected, restoring the complete double helix. Untangling of helices is more complex. First, a topoisomerase breaks both strands of the DNA helix and physically connects them to tyrosines on its surface, ensuring that the ends do not float away from each other. Next, a tangled DNA strand is passed through the break, resolving the knot. Finally, the original strands are reconnected, forming the original, continuous double helix. The two topoisomerases pictured here perform these untangling operations. Of course, these operations must be performed with the utmost delicacy. A single mistake will corrupt the genetic information stored in the DNA strands.

Topoisomerases

DNA Topoisomerase I

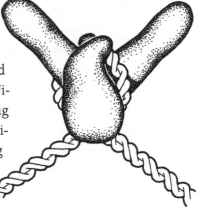

DNA Gyrase

DNA Polymerase

DNA polymerase is our most accurate enzyme. It gently separates the two strands of a DNA double helix and builds a new mate for each strand, copying the hereditary plans passed from parent to child. The result is two identical double helices, exact copies of the original, each with one strand from the original double helix and one new strand. DNA polymerase builds the new strands to exactly complement the originals. Cytosine on the old strand is always paired with guanine on the new strand, and vice versa. Adenine is always paired with thymine, and thymine with adenine. The strong chemical preference of adenine for thymine, and of cytosine for guanine, provides the language with which we duplicate and utilize the information held in DNA strands.

If DNA polymerase makes a mistake, perhaps by placing a guanine, not a thymine, next to an adenine at one position, the result is a *mutation*. This mistyped

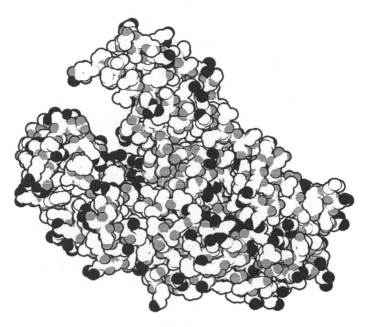

DNA Polymerase

character in the genetic blueprint may specify a change in a protein, causing a improper amino acid to be added when the protein is constructed. Most mutations are transparent, making a change that only marginally modifies the protein. Some changes, however, are dangerous or even lethal: if the sixth amino acid in hemoglobin (page 75) is changed to valine, the protein will form unnatural, ropy fibers that interfere with the circulation of red blood cells, causing sickle cell anemia. Or the mutation may corrupt important control information, causing proteins to be made when they are not needed.

Three separate mechanisms ensure fidelity in copying DNA, to minimize the risk of mutation. First and foremost is the chemical selectivity of base pairing: adenine with thymine, guanine with cytosine. The strength of this specificity is strong enough to ensure that a mistake is made only once in every 100,000 bases. On top of this, DNA polymerase also contains a digestive capability, which is sensitive to mistakes. If a mispaired base is added, it will be quickly digested away, allowing the proper base to be added in its place. DNA polymerase adds bases, then wiggles them to check if they are right, improving the accuracy one hundredfold. Finally, a separate repair enzyme searches the new DNA strands for mistakes, after the polymerase finishes. The final accuracy is about one mistake per billion nucleotides, or half a dozen mistakes each cell division. This error rate is amazing, and far better than any information system in our computerized world. One mistake per billion is roughly equivalent to copying a thousand books and making only one mistake.

Ironically, mutations are essential to the evolution of life. Were it not for an occasional mistake in the copying of DNA, the Earth might now hold only a thin film of bacterial life. Nature relies on mutations to progress. Occasionally, a mutation alters an organism in an entirely new way. A minor change in a protein controlling embryonic development may allow a leg to grow a little longer. The taller giraffe will then be able to reach higher leaves, or the sleeker cheetah to run faster. A digestive protein may be changed so that it now destroys a potent toxin, so koala bears can eat eucalyptus or monarch butterfly caterpillars can eat milkweed leaves. By these small fits and starts, new species are born. If they are sufficiently successful, they will thrive in their new environmental niche.

Bacterial DNA polymerases have found wide use in the biotechnology industry. (Our polymerases are typically larger than the simple bacterial polymerase illustrated here.) These small enzymes are used to copy DNA artificially in a test tube. The *polymerase chain reaction* (PCR), a clever method for amplifying minute quantities of DNA into useful amounts, relies on the action of these enzymes. PCR has found use in many glamorous fields. The tiny amounts of DNA from a blood stain may be copied and amplified, and compared with the DNA of a suspected criminal. This process, now being tested in the courts, is termed *DNA fingerprinting*. Ancient DNA samples may also be copied and amplified by PCR to look at the pedigree of a mummy or, in fiction, to build a dinosaur.

Messenger RNA

Proteins are not built directly from the information carried in strands of DNA. Instead, an intermediary molecule delivers the genetic information from the DNA library to the machinery of protein synthesis. The messenger carries the information in the same language as the DNA, encoded in a string of nucleic acid bases. But the messenger is made of RNA, instead of DNA, and is composed of a single strand, instead of the sheltered double helix of DNA. A messenger RNA is transcribed from DNA, used to build dozens or thousands of copies of a protein molecule, and then thrown away. Each strand of our messenger RNA typically contains one *gene*, the instructions to build a single protein (bacteria often make longer messenger RNA molecules, containing the information for several proteins with related functions). At the start of the message, a guanine nucleotide is attached, backwards. This reversed nucleotide protects the end from degradation—there are many RNA-digesting enzymes in the cytoplasm—and provides a handy signal to the protein synthetic machinery, specifying the proper end on which to start. The tail end of the message typically contains a string of about 200 adenine nucleotides. This polyadenine tail protects the end from degradation, at least for as long as it lasts.

The use of an expendable message allows many opportunities to regulate the reading and use of our genetic information. Most regions of DNA are held in storage—messenger RNA is made only when a particular piece of information is needed. A host of DNA-reading proteins (page 45) block or enhance the construction of messenger RNA, controlling the flow of information from DNA to messenger RNA. Once a message is made, it must be transported to the cytoplasm, where proteins are made according to its instructions. This transport of the message from nucleus to cytoplasm is carefully controlled by the nuclear pore (page 88). Once the message is in the cytoplasm, one or one thousand proteins can be built from it, depending on its life span, which may be minutes or hours. The long adenine tails of messenger RNA are slowly degraded. When the tail is shorter than about 30 adenines, the entire RNA is destroyed. On some important messages, the tail is continually repaired, extending their lives. Other messages are specifically attacked by proteins in the cytoplasm, leaving time to construct only a

Messenger RNA

handful of proteins. In addition, a single strand of RNA is much more flexible than the stoical double helix of DNA, so the shape of the messenger RNA may also be used to regulate the number of proteins being built. Most of an RNA message lies in formless coils, but small regions may be designed to form stable hairpin loops. These kinks stall the protein synthesis machinery, slowing protein production.

Surprisingly, our genes are not written in one continuous stretch in our DNA. In most, the plans for building a protein are interrupted by *introns*, stretches of DNA that do not specify amino acids. This is like reading a sentence /Nature is what we know/ with interspersed text that must be removed /Yet have no art to say/ before it makes any sense. Just as the lines of Emily Dickinson interspersed here are composed of the same letters as the text, the introns in DNA are composed of the same four bases used to write the instructions for a protein. Unlike the sentence, however, the introns may not themselves "mean" anything. Perhaps they are evolutionary fossils, left over from billions of years of evolutionary shuffling.

Soon after messenger RNA is transcribed from DNA, it is processed to remove any introns, piecing together the complete instructions for building a protein. *Small nuclear ribonucleoproteins* (snRNPs), composed of both RNA and protein, perform the editorial functions. The RNA component, since it is written in the language of nucleic acids, reads and recognizes the introns to be excised. At any time, several snRNP will be working on a given RNA message, often grouping together into a large complex appropriately termed a *splicosome*. Inside, the extra bits are clipped out and the mature messenger RNA is pieced together. Remarkably, some RNA strands can perform this reaction without the help of an enzyme. The RNA itself performs the cutting and pasting. Such self-splicing RNA molecules have been termed *ribozymes*.

RNA Polymerase

RNA polymerase transcribes our genetic information, unwinding the DNA double helix and constructing an RNA strand exactly complementary to the sequence of bases stored therein. If the resulting RNA is a messenger RNA, other enzymes add a guanine cap at the beginning and attach a long string of adenines at the end. If it is a transfer RNA (page 47), nuclear enzymes clip it to size and chemically modify certain nucleotides, allowing it to fold into its unusual shape. If it is ribosomal RNA, it is cut to size and combined with proteins to build a ribosome (page 49).

RNA Polymerases

RNA polymerase is not nearly as careful as DNA polymerase (page 40). Errors in RNA are evanescent and are not passed on to future generations like errors in DNA. As RNA polymerase adds nucleotides one by one to a growing RNA strand, it may make an error every 10,000 bases, or one in every other strand of RNA. This error rate is not a great problem, however, as a similar error rate would be for DNA polymerase. Most mutations are relatively harmless: they may change a single amino acid in a protein, which will most often be in an inconsequential position. The RNA strand is only temporary, so only a few modified proteins will be made. In rare cases, if these proteins don't function, another messenger RNA may be constructed and the proteins replaced. But DNA polymerase must be far more careful than this, because any error it makes will be passed from cell to cell and from parent to child, never to be corrected.

I

II

RNA polymerase, because of its vital function, is an easy target. Poisons that disrupt protein synthesis are often very potent, because protein synthesis is essential to life. But they may be unusually slow-acting, in spite of their inevitable lethality, because individual cells generally carry sufficient protein to last hours or days. Often, poisons of protein synthesis cause death through extensive damage to the liver, our major site of protein production. α-*Amanitin* from the death cap mushroom is a well-known example. Eating even one of these mushrooms leads to coma and death in several days, as the poison slowly spreads through the body attacking RNA polymerase.

Sequence-Specific DNA-Binding Proteins

Our DNA contains information for building about 60,000 proteins. Some, such as the enzymes that harness chemical energy and the proteins that form the cytoskeleton, are needed in every cell. Many others, however, are needed only by certain types of cells. Red blood cells need to make hemoglobin (page 75) in large quantities; most other cells do not wish to make it. Muscle cells need far more actin (page 90) and myosin (page 105) than other cells. Other proteins are needed only at certain times. Antibodies (page 124) are made only when an infection is being fought; tyrosinase (page 104) is made after we spend time in the sun. Thousands of proteins are used only in the first nine months of life, to direct the growing, folding, stretching cells of an embryo. If these proteins are made later in life, unnatural growth will occur, leading to forms of cancer.

The synthesis of proteins must be controlled, otherwise, cells would make every possible protein at the same time. The easiest way to regulate the reading of genetic information—ensuring that each protein is made in the proper place and at the proper time—is to physically block the DNA. *Repressor* proteins do just this: they clamp onto the DNA at precisely the spot where RNA polymerase starts to build an RNA message, forming a roadblock that halts the enzyme. Repressors seek out specific sequences of bases and bind tightly to them. The simplest examples have been studied in bacteria. The bacterial *met repressor* binds to DNA with the sequence -A-G-A-C-G-T-C-T-, which is found just before the region containing the information needed to build the enzymes of methionine synthesis. When methionine is needed, met repressor releases the DNA and the appropriate enzymes are ultimately made. The opposite approach is taken by *activators*. Instead of physically blocking the segment of DNA, activators coax RNA polymerase into place, enhancing the synthesis of a given RNA message. The bacterial *catabolite gene activator protein*, which binds to long DNA sequences such as -T-G-T-G-

Sequence-Specific DNA-Binding Proteins

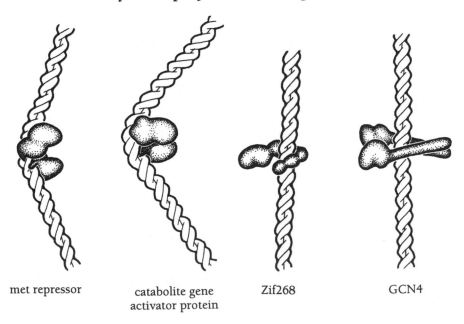

met repressor   catabolite gene   Zif268   GCN4
activator protein

A-G-T-T-A-G-C-T-C-A-C-T-, is an example. It senses increased levels of cyclic AMP (see page 151) and increases the synthesis of enzymes involved in energy production. Thousands of repressors, activators, enhancers, and other gene regulatory proteins are dotted along our strands of DNA, controlling which genes are used in blood cells and which in muscle, which are used every day and which never again after birth.

Regulatory proteins, unlike the polymerases, typically do not unwind DNA. Instead they have short arms that hug the grooves of the double helix, delicately feeling the exposed edges of the DNA bases. The DNA-binding proteins pictured here have distinct designs for reading the sequence of DNA. The catabolite gene activator protein sticks an elbow of protein into the groove of DNA, reading specific bases. The met repressor uses an extended segment of protein chain, lying flat against the base of the DNA groove, to perform the recognition. Some proteins regulating development, such as Zif268, use an atom of zinc to form a rigid DNA-probing module, termed a "zinc finger." Other regulatory proteins, such as GCN4, contain a chain with every seventh amino acid a leucine, termed a "leucine zipper," that forms a structure similar to a clothespin pinching the DNA helix.

Transfer RNA

Transfer RNA translates the nucleotide language of messenger RNA into the amino acid language of a protein. Three successive nucleotides in an RNA message translate to one amino acid. Each triplet of bases is termed a *codon*; each different codon specifies a single amino acid. C-U-G translates as leucine, C-G-G as arginine, and so on. The three codons U-A-A, U-G-A, and U-A-G are translated as STOP! For many years, scientists (and philosophers) have searched for some relationship between the nature of each of the 20 amino acids and the nature of the three bases in each codon. These attempts have been uniformly unsuccessful. The only relationship appears to be the code used by transfer RNA, the Rosetta stone that translates nucleic acid codons into protein amino acids.

Cells make a battery of L-shaped transfer RNA molecules for reading the 20 different amino acids. Each is a short chain of 70 to 90 nucleotides folded into a trefoil loop. The two ends are close together at the pointed leg of the L (at the top in the illustration), the longer of which carries the amino acid. The center of the chain forms a loop at the rounded leg of the L (at the bottom in

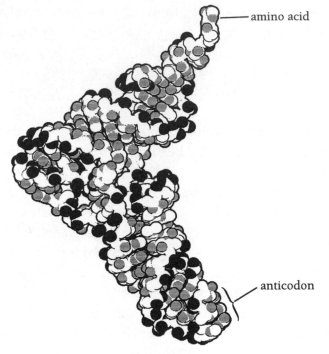

amino acid

anticodon

Transfer RNA

the illustration), exposing three nucleotides that read the
triplet code on messenger RNA, termed the *anticodon*.
The other two loops of the trefoil are bundled into the
elbow, providing structure to the entire molecule. The
four normal RNA bases—adenine, uracil, guanine, and
cytosine—do not seem to provide enough structural lati-
tude for designing this complex structure, so many of the
bases in transfer RNA are chemically modified to enhance
their interactions.

  Of course, each type of transfer RNA must be carefully matched with its proper amino acid. A collection of
highly specific enzymes, the *aminoacyl-tRNA synthetases*, carry
this responsibility. They attach an amino acid to the tip of the
proper transfer RNA. Each of these enzymes is different—a
different one is built for each type of transfer RNA. Some are

Aminoacyl-tRNA Synthetases

serine

glutamate

48

small, others may be five times larger. Some are formed of a single protein chain, others of two or four identical chains, others of several different chains. Some recognize the three bases of the anticodon, rejecting all but the proper transfer RNA. Surprisingly, others recognize bases in the elbow of the transfer RNA, ignoring the anticodon entirely.

Ribosomes

Ribosomes carry the formidable task of constructing new proteins. Ribosomes start by sliding down a messenger RNA until the codon A-U-G is found. A-U-G has two meanings: it is the codon specifying methionine, and it is the message START. When the ribosome finds the first A-U-G, it pairs a methionine transfer RNA with it, and the protein has been started with its first amino acid. The ribosome then steps three nucleotides forward on the messenger RNA and aligns the proper transfer RNA with the next codon. This second amino acid is then added to the growing protein. Similarly, the ribosome steps along a messenger RNA, three bases at a time. At each position, it aligns the proper transfer RNA and adds its amino acid to the protein chain. At a speed of 10 amino acids per second, a ribosome takes only a minute to make a typical protein, while at the same time remaining remarkably accurate. Ribosomes may make a single error in 10,000 amino acids, or once in about two dozen proteins. Notice that every new protein begins with methionine, because of the dual role played by the codon A-U-G. This may not suit the function of every protein, so a specialized digestive enzyme clips methionine off newly made proteins.

Ribosomes are large, asymmetric factories, formed of four strands of RNA and more than 80 proteins. Experiments have shown that many of the proteins are not absolutely essential for proper function—it is thought that the RNA performs much of the synthetic work. Our mitochondria contain smaller ribosomes than those in the rest of the cell, composed of three RNA strands and about 55 proteins. These smaller ribosomes are similar to the small ribosomes found in bacteria. This surprising fact, combined with many other lines of compelling evidence, has led to the proposal that our mitochondria are cellular fossils. It is thought that a bacterium chose to live inside an animal cell in the distant past. As the two evolved toward the present, they divided the tasks of living: the bacterium inside specialized in energy production, the surrounding cell specialized in gathering and digesting food. Today, the pair cannot live without one another. The bacteria are now our mitochondria,

Ribosomes

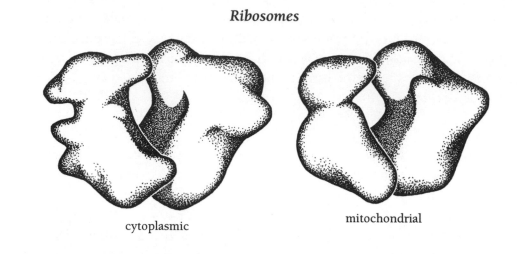

cytoplasmic   mitochondrial

comfortably living and dividing in each of our cells. They retain some of their autonomy, however. They carry their own DNA and their own ribosomes, smaller than those in the rest of the cell. At conception, the mother's egg cell provides the initial inoculation that leads to a body-wide infection of mitochondria.

Ribosomes are essential to life, so they are often targets for poisons and drugs. The difference between a poison and a drug is the identity of the victim. If our own ribosomes are attacked, the compound is a poison; if the smaller ribosomes of bacteria fall victim, the compound will be useful as an antibiotic drug. *Chloramphenicol*, *tetracycline*, and *streptomycin* fall in the latter category, attacking bacterial ribosomes and leaving our larger ribosomes unharmed. Unfortunately, they may also attack the smaller ribosomes in our mitochondria, leading to unwelcome side effects. Diphtheria toxin (page 113), one of the deadliest toxins known, attacks our own machinery, blocking the action of a protein factor that assists ribosomes in their task.

Molecular Chaperones

Proteins are made as long, formless chains. After they are released from a ribosome, many proteins fold spontaneously into their proper shape. Others, however, require some assistance. Normally, proteins fold into a compact ball, sheltering leucines and phenylalanines inside and displaying charged amino acids, which interact more favorably with water, on their surface. But the same result may be achieved by bringing several different protein chains together, with all of their carbon-rich amino acids arranged side by side. The result would be a useless, gluey aggregate. *Molecular chaperones* prevent this from happening. Inside their large, comfortable cavities, large enough for only a single protein chain, a juvenile protein chain is guided down the proper folding pathway, safe from the influence of other maturing protein chains.

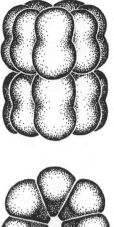

Two additional problems must be addressed in the folding of proteins: cysteines must be paired into the proper sulfur–sulfur linkages, and each proline must be coaxed into its proper shape. The former problem is solved by *disulfide bond formation protein*. It contains a very reactive sulfur–sulfur bond on its surface. Upon finding two sulfur atoms in close proximity on a folding protein, the reactive bond breaks, forming a new sulfur–sulfur bond in the maturing protein. This process is typically the slowest aspect of protein folding, as a protein searches each possibility for the proper pairing

Molecular Chaperone

of cysteine to cysteine. The second problem occurs with the amino acid proline, which forms a rigid kink in protein chains. This kink is useful for turning tight corners and fitting into odd places, and thus is essential for proper folding. But proline can adopt two forms, termed *cis* and *trans*, with different shapes: *cis* is sharply kinked, *trans* less so. Each proline in a protein will need to be one or the

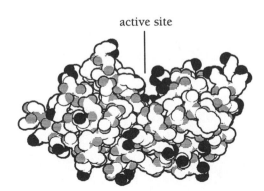

active site

Disulfide Bond Formation Protein

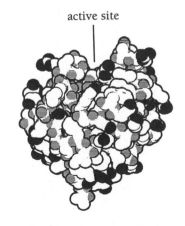

active site

Proline **cis-trans** *Isomerase*

51

other—if any are in the wrong form, the protein will not fold properly. The enzyme *proline cis-trans isomerase* converts one to the other. Because of this critical function, it is a common enzyme found in nearly every type of cell. It is one of the smallest enzymes, containing only 165 amino acids. The active site is formed in a shallow pocket on one side.

Ubiquitin

In the warm, salty environment of our cells, most proteins survive for hours or days. In special cases, such as the crystallins (page 155) in our eye lenses, proteins may last our entire life. Others, however, have a lifetime of mere minutes. These include regulatory and control proteins that guide the minute-by-minute actions of the cell, such as *cyclin*, which guides the processes of cell division. Enzymes that perform a key metabolic step may also have very short lifetimes. For instance, the enzyme *ornithine decarboxylase*, which begins the synthesis of spermine (a charged molecule that helps stabilize nucleic acids), is built only when spermine is needed and is destroyed after about half an hour, when supplies are sufficient. These proteins are built exactly when they are needed and destroyed soon after. It may seem wasteful to destroy a protein minutes after its synthesis. This is, however, a particularly direct method of controlling the action of a protein, placing the control squarely upon the genetic code. There is no need to devise additional molecules to turn the protein on or off. When a cell is done with the protein, it is destroyed; when it is needed again, a new one is built.

Ubiquitin provides the signal to destroy an obsolete protein. As its name implies, it is found in nearly all living things. Ubiquitin has changed very little over the evolution of life, underscoring its key role. Three enzymes attach strings of ubiquitin to lysines on an obsolete protein. The actual method by which they recognize these obsolete proteins is still a matter of controversy. It is thought that the presence of lysine and arginine at the end of the chain may identify proteins destined for a short life. A stretch rich in proline, glutamic acid, serine, and threonine is also thought to be a signal to the ubiquitin destruction machinery.

Of course, obsolete proteins cannot be destroyed by small protein-digesting enzymes like chymotrypsin (page 21). Small proteases cannot be released inside cells: they are too small and stable, they move quickly, and they

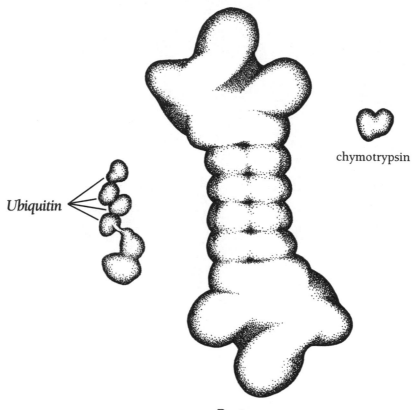

Proteosome

act indiscriminately. The protein-cutting enzymes acting inside the cell, by con-trast, are large complexes of many enzymes, termed *proteosomes*. They carefully search for ubiquitin tags, digesting only those proteins specifically scheduled for destruction.

3
Powering the Body

F ood is composed of hundreds of different molecules. Indeed, molecules similar to most of those pictured in this book appear on our dinner plate at one time or another. Meats and vegetables are rich in protein and nucleic acid. Potatoes and corn are rich in starchy carbohydrate, and fruits are naturally sweetened with simple sugars. A specialized cadre of enzymes, secreted into the stomach and intestines, break these diverse food molecules into manageable pieces. Most of these enzymes are *hydrolases*, which use individual water molecules to break the bonds in their target molecules. Proteins are broken into smaller pieces, then into individual amino acids. Nucleic acids are broken into nucleotides and often further separated into their component phosphates, sugars, and bases. Polysaccharides yield sweet simple sugars.

These pieces may be used directly as building blocks for new protein, carbohydrate, and nucleic acid, providing the raw materials to build new muscle, repair injuries, and rejuvenate blood. But we also require a great deal of food simply for energy. Food molecules are burned as fuel, like gasoline in an automobile. The energy powers the motion of our arms and legs, contracting and releasing our muscles. Energy from the burning of food heats the body, maintaining optimal warmth; enzymes work efficiently only at a narrow range of temperatures. And perhaps of greatest importance, energy is consumed by enzymes as they perform their diverse chemical reactions. Energy drives each reaction, ensuring that only the proper chemical modification is performed, at only the time it is needed.

The central source of biological energy on Earth is sunlight. Plants capture red and violet light (leaving only the greens to color the leaves) and use it to convert carbon dioxide into sugar. This sugar provides the energy that powers the living world. We begin our utilization of energy after all of the hard work is finished. All that remains is for us to eat the plants, sort out the pieces that we need for building, and burn the rest for energy.

Digestion

As we sit at the table eating our supper, digestive enzymes are already hard at work, breaking individual molecules into manageable pieces. Digestion begins immediately, even before we swallow. Saliva contains *amylases* that begin to break carbohydrates into simple sugars. Digestion then begins in earnest when food reaches the stomach. Hydrochloric acid is added, unfolding the tough knots of proteins, and making them more accessible to acid-loving *proteinases*. Food then passes into the intestine, where the acid is neutralized by bicarbonate. The pancreas then adds a host of enzymes—*lipases*, *nucleases*, more proteinases—which cut everything into small pieces. The final steps of cleavage are performed by dozens of specific enzymes tethered tightly to the walls of the intestine. The resulting sugars, nucleotides, amino acids, and fatty acids are absorbed by the feathery cells lining the intestine and delivered to the bloodstream for distribution to hungry cells throughout the body.

Serine Proteinases

The synthesis of protein-digesting enzymes is a delicate business. The cells lining the digestive tract must make these enzymes and deliver them to the stomach and intestines without digesting themselves in the process. To do this, they build digestive enzymes in an inactive form, termed *proenzymes* or *zymogens*, that are activated only after they are safely delivered out of the cell. These harmless proenzymes are often activated by cleavage. In the case of chymotrypsin (page 21), two small loops on the protein surface are the pin to the grenade. When they are removed, the enzyme relaxes slightly, bringing the amino acids of the active site into optimal destructive alignment.

Serine Proteinases

active site

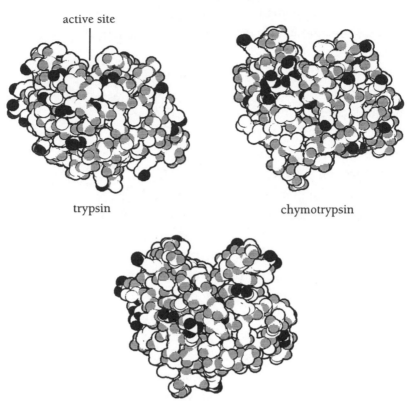

trypsin

chymotrypsin

elastase

Serine proteinases such as *trypsin*, *chymotrypsin*, and *elastase* are secreted by the pancreas into the intestine. They are all similar in shape and composition, and all use a serine amino acid to perform their digestive reaction (see chymotrypsin, page 21). Because of their similarity, they are thought to have evolved from a common ancestor. All are characteristically bean-shaped, with a cleft on one side that holds the active site. Small differences in the shape of the active site give each of these proteinases their specific tastes for proteins: trypsin cleaves protein chains next to positively charged amino acids, chymotrypsin next to bulky amino acids, and *elastase* next to small amino acids. Together, a mixture of proteinases can chop proteins into short pieces of about a dozen amino acids.

As digestion reaches completion, we need some method of halting these destructive enzymes. Several *serpin* proteins carry out this task, inactivating pro-

teinases after their job is done. (This almost
mythologically evocative name has a rather
mundane root: *serine proteinase inhibitor*).
Serpins lock into the active site of proteinases
and completely block their action. The perfect
fit of serpin to proteinase ensures that the
serpin is not itself chopped into pieces.
Trypsin inhibitor is an example of an intestinal proteinase inhibitor. A number of
foods, such as eggs and potatoes, also contain natural proteinase inhibitors, designed to
protect their stores of protein from bacterial digestive enzymes.

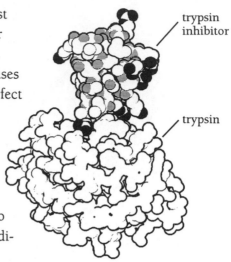

trypsin
inhibitor

trypsin

 Bacteria and fungi are often much smaller
than the food they are eating, so they typically secrete their serine proteinases
outside their bodies, digesting food in the local area. Their proteinases are thought
to have evolved separately from their human intestinal counterparts. But in spite
of their different origin and significantly different structure, their active site machinery is virtually identical to that of the intestinal enzymes. This is a molecular example of convergent evolution: two organisms developing the same solution to a common problem, much as the wings of bats and the wings of birds
evolved from different flightless ancestors.

 A similar class of proteinases, termed *cysteine proteinases*, use the sulfur
atom in a cysteine amino acid to perform a cleavage reaction, in place of the
oxygen atom in a serine. Tiny digestive compartments inside each of our cells,
termed *lysosomes*, contain the cysteine proteinases *cathepsin B* and *cathepsin H*, as
well as dozens of other digestive enzymes. Inside lysosomes, old proteins are
digested and their amino acids are recycled to build new proteins. Two cysteine
proteinases made by plants may be found in the local grocery store. *Papain* is
found in the milky latex of papaya plants, where it helps to protect the plant from
herbivores. It is commonly used in the enzymatic cleaning solutions for contact
lenses. Papain digests the cloudy deposit of tear proteins deposited on the lenses.
Actinidin is found in the kiwifruit (Chinese gooseberry). Because of this proteinase, kiwifruit-flavored gelatin desserts will often be a failure. Actinidin quickly
digests gelatin, which is almost pure collagen (page 95), so it never gels.

Aspartyl Proteinases

Aspartyl proteinases use a pair of aspartate amino acids to cleave proteins. They are typically active only under very acidic conditions, making them the ideal choice for use in the stomach, which is normally filled with hydrochloric acid. *Pepsin* is the major digestive enzyme active in the stomach. Pepsin played a central role in early studies of biochemistry, when arguments raged over the very nature of enzymes. Many scientists did not believe that enzymes themselves performed chemical cleavages; instead, they thought that enzymes merely carried small reactive molecules that did all the work. The choice of a digestive enzyme for these studies is not a surprise. Digestive enzymes are abundant and extraordinarily stable. They may be easily coaxed to crystallize, ensuring purity. They also cleave a variety of simple model compounds, simplifying the gathering of data.

Aspartyl Proteinases

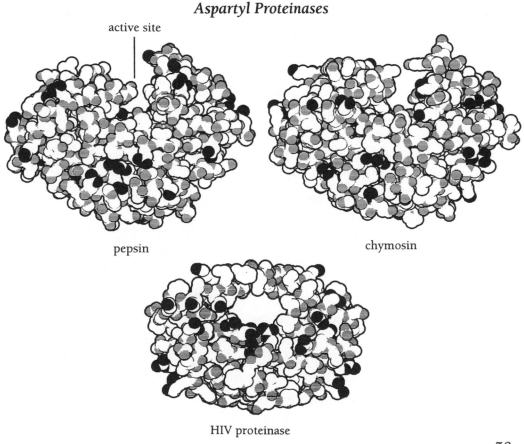

active site

pepsin

chymosin

HIV proteinase

Chymosin, also known as rennin, is a similar aspartyl proteinase found in cattle. It is traditionally used in the manufacture of cheese. A chymosin-rich extract of calf stomach, called rennet, curdles the proteins in milk. Today, however, aspartyl proteinases from bacteria or fungi, such as *penicillopepsin* from *Penicillium* mold, are gaining wider usage. Since they are secreted by fungi, they are easier to isolate in pure form than the enzymes from livestock.

HIV proteinase is an aspartyl proteinase made by the AIDS virus. It is structurally different from pepsin and chymosin, being smaller and composed of two subunits instead of one. Two flexible "flaps" cover the active site, entrapping the target protein chain inside an active-site tunnel. HIV proteinase is not used for random digestion. Instead, it makes a few specific cuts necessary in the life cycle of the virus. HIV proteins are not made individually, like normal human proteins: they are made in the form of a long *polyprotein* chain. HIV protease clips these long chains in a few specific places, releasing the active proteins, which then assemble to form new, infectious viruses. HIV proteinase, since it performs an essential role in the reproduction of HIV, is a major target for drug design in the ongoing fight against AIDS.

Peptidases

Our digestive proteinases break proteins into pieces of about a dozen amino acids in length, termed *peptides*, but are unable to cleave them further. They simply cannot get a firm grip on these tiny fragments. So, peptides are digested by a second set of digestive enzymes, termed *peptidases*. Peptidases perform the final steps of protein digestion by clipping from the ends of these small peptides, releasing amino acids one at a time. Aminopeptidases start at one end of a peptide, the end with a free a-mine, and carboxypeptidase starts at the opposite end, that with a free acid.

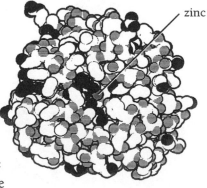

zinc

Carboxypeptidase requires an atom of zinc for its digestive action. The active site is located in a small pocket (at the front in the illustration), which is lined with reactive amino acids. The zinc is held on one side of the pocket, where it is thought to weaken the bond

Carboxypeptidase

Peptidases

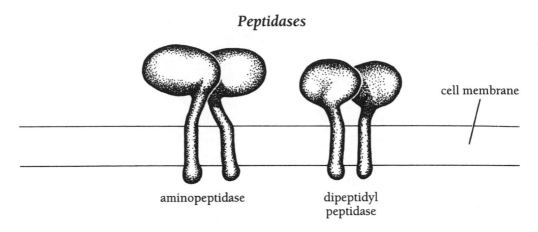

cell membrane

aminopeptidase

dipeptidyl
peptidase

being broken. The terminal amino acid of a peptide fits perfectly into this pocket and is snipped off. Like the proteinases, carboxypeptidase is a small, sturdy enzyme, which allows it to seek out and attack peptides under the harsh conditions of digestion.

Aminopeptidases are often found tethered to the inner surface of the intestine, extending into the digestive tract. Unlike the soluble proteinases such as pepsin and trypsin, the forest of peptidases bristling from intestinal cells do not have to be remade for every meal. They have a long tail composed of carbon-rich amino acids, which spans the cell membrane, locking the enzymes in place (for more on cell membranes, see page 82). The large, globular heads contain the peptide-cutting active sites. *Dipeptidyl peptidases* also extend from the walls of the intestine, breaking dipeptides (peptides containing only two amino acids) in half.

Amylases

Two forms of digestible carbohydrate are common in a typical diet: starch from vegetables and glycogen from meat (vegetables also contain cellulose, which is indigestible but plays an important role as roughage). Both are long chains of the simple sugar glucose, connected in a slightly different manner. These long chains are broken into manageable pieces by *amylases*. Saliva contains amylases that begin digestion as we chew, and a second amylase produced by the pancreas finishes the job. Both of these amylases require calcium for action. Amylase cleaves starch and glycogen in the center of the chains. It is a typical digestive enzyme: small in size, which allows it to search rapidly for targets, and sturdy in design,

which allows it to survive the harsh environment of the digestive system. *Glucoamylase*, on the other hand, removes sugars
one at a time from the end of a starch chain. It is composed of
two globular elements connected by a flexible linker. The
smaller half anchors glucoamylase tightly to the surface of a
starch grain, and the larger half breaks glucose molecules
one at a time off of nearby starch chains. The tether linking the two ends is kept supple by short, knobby sugar
chains attached along its length.

Amylase

Glucoamylase

    Natural sources of simple sugars are relatively rare.
Simple sugars are typically produced in fruits and nectars
only when they are needed to entice an animal to distribute
pollen or seeds. Sugars are normally stored in more compact forms—starch and
glycogen—which take far less room and soak up far less water. Sweet simple
sugars are in great demand, however, in the food industry. Much of the sugar used
to sweeten commercial foods is produced with amylases purified from fungi. Fungi
are excellent sources for amylases, because they secrete the enzymes into the food
they are growing upon, making the enzymes easy to collect and purify. After a
short exposure to purified amylase, powdery corn starch is converted to sweet
corn syrup, which is a mixture of shorter, sweeter carbohydrate chains. A more
intensive application of amylase digests corn starch completely into pure glucose.
The taste of glucose, however, is not pleasant to most people, so it is converted to
sweeter fructose by other purified enzymes, such as d-xylose isomerase (page 34).

Disaccharidases

The final steps of carbohydrate digestion are performed by disaccharidases, which
cleave disaccharides (chains of two sugars) in half. Two simple sugars are released
and quickly absorbed by intestinal cells. Like the peptidases of protein digestion,
disaccharidases are tethered to the walls of the intestine. The two disaccharidases
pictured here each contain two distinct enzymes, connected to the same cell membrane anchor. *Maltase-glucoamylase* and *sucrase-isomaltase* break various combinations of fructose and glucose disaccharides into usable fructose and glucose.

    Many people are subjected to severe inconvenience by a deficiency of
lactase, a disaccharidase that digests lactose in milk products. Lactose is normally
broken into the simple sugars glucose and galactose in the intestine. But lactose-

Disaccharidases

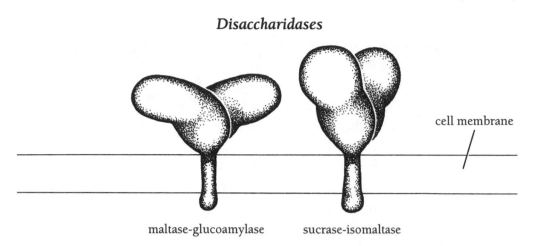

cell membrane

maltase-glucoamylase   sucrase-isomaltase

intolerant individuals do not make this enzyme. Undigested lactose remains in their digestive system, where it is fermented by bacteria, often releasing gas bubbles in the process. This may lead to painful cramps. For individuals who have a craving for ice cream but are cursed with a congenital inability to digest it, substitutes for the enzyme are now available. One can eat the dairy product and the enzyme to digest it at the same time.

Lipases

Triglycerides, commonly known as *fats*, are the typical lipids encountered in our diet (for more on lipids, see page 82). They are composed of three carbon-rich tails connected to a small molecule of glycerol. These three tails render triglycerides virtually insoluble in water. Triglycerides are used primarily for storage of energy. Because they shun water, they are a very compact molecule for storage, forming oily droplets inside cells. In cold climates, cells filled with droplets of triglyceride are also used for insulation.

Lipases are assigned the difficult task of digesting fats and oils. Proteinases and amylases simply drift up to their targets and begin work, but lipases must attack a target that is insoluble in water. A droplet of lipid presents only a sheer, oily surface to the surrounding water, with nothing protruding for a digestive enzyme to clip off. To deal with this problem, lipases have evolved a unique mode of action. Like the proteinases, they are small and stable. They tend to be disk-shaped, with a pocket on one face, often covered by a flexible flap. When a

lipase finds a droplet of fat or oil, it sticks tight
against the flat surface. It squeezes out all of the
water in between, and the flap opens to reveal
the active site machinery. Lipids are extracted
one at a time and digested. The enzyme sticks
like a limpet shell on a tide pool rock, grazing
along a droplet of fat.

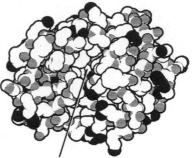

active site

Small lipases are also a major component
of the venom of cobras and rattlesnakes, and in the
sting of bees. Small phospholipases, similar to the
lipases we use for digestion, attack cellular barriers around the bite or sting, bursting cells
and allowing the venom to insinuate further
into the unfortunate victim.

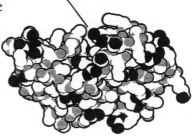

Serum Lipoproteins

Many fats and lipids are not immediately digested
but instead are stored for future use or used to con-

Phospholipase

struct new cell membranes (page 82). The delivery of lipids poses something of a
problem, as the blood is a watery environment, and greasy fats and water do not
mix. We solve this dilemma by shipping them inside protein supertankers termed
serum lipoproteins, composed of a droplet of fat sheltered by a shell of protein. The
proteins play two roles: they separate the greasy contents from water and they
carry messages on their surfaces, signaling to fat cells to hoist in their tasty cargo.

Serum lipoproteins are built in many sizes for transporting different fats
and lipids. *Chylomicrons,* filled with triglyceride fats from the diet, are the largest.
(Chylomicrons are not illustrated here; at 2 million times magnification, each one
would measure some two meters in diameter.) These huge supertankers are
launched into the blood and are slowly digested by *lipoprotein lipase*, which extracts fat molecules one by one and severs their carbon-rich tails, making them
available for further digestion. Surprisingly, this enzyme is found tethered inside
the walls of blood vessels—digestion of fat begins in the blood, as it is being
transported to hungry cells. The *low-density lipoproteins* (LDL) are smaller than
chylomicrons (about the size of a ribosome) and primarily carry cholesterol. They
are launched from the liver and are picked up by cells throughout the body, which

use the cholesterol for building membranes or constructing hormones. The *high-density lipoproteins* (HDL) are even smaller (and denser because they are richer in protein). When newly made, HDL are disk-shaped, with a ring of protein surrounding a phospholipid bilayer. As they travel through the blood, they soak up cholesterol like a sponge, swelling to a spherical shape. Ultimately, HDL deliver excess cholesterol back to the liver.

Serum Lipoproteins

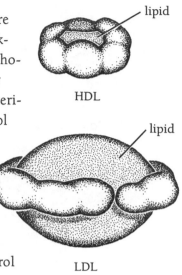

HDL

LDL

The mismanagement of lipid transport can be disastrous. Cholesterol is transported primarily in LDL. A consistently high level of LDL circulating in the blood, often the result of a diet high in cholesterol, has been shown to lead to atherosclerosis. The high levels of cholesterol gradually help form hard plaques inside blood vessels, restricting normal blood flow. This is particularly dangerous in the blood vessels that feed the heart, which requires a constant supply of oxygen and sugar. High levels of HDL have the opposite effect, cleaning up excess cholesterol and reducing the risk of coronary heart disease.

Nucleases

Our nucleases are typical of digestive enzymes: they are small and tough. They are relatively insensitive to the actual sequence of nucleotides of the nucleic acids that they digest. They make cuts wherever they happen to land, ignoring entirely what the strands "mean." *Deoxyribonuclease* cleaves the double helix of DNA, breaking both strands. Its active site machinery lies in a large depression on one side, shaped to fit against the helical shape of DNA. This shallow depression is better suited to the wide double helix than is the deep groove typical of proteinases. *Ribonuclease*, on the other hand, breaks narrow single strands of RNA into small pieces. As with the proteinases, its active site lies in a comfortable groove, just wide enough to fit a strand of RNA.

Information can be a dangerous commodity, making nucleases a necessity. Viruses attack by injecting foreign information into our cells. The viral nucleic acid, which may be RNA or DNA, depending on the type, disrupts the normal processes of

Nucleases

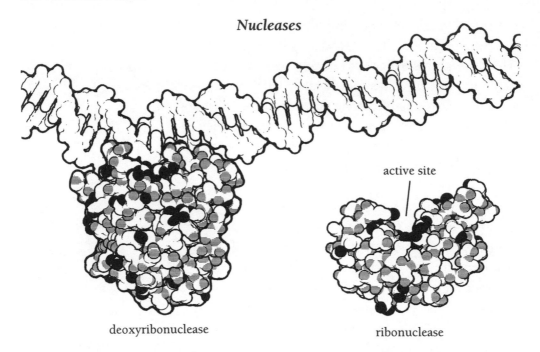

deoxyribonuclease   ribonuclease

an infected cell and forces it to make thousands of new viruses. Ribonuclease forms our first line of defense against this subversive literature. It circulates in the blood, in sweat, and in tears, destroying viral RNA before it has a chance to attack.

Bacteria, like us, are under constant siege by viruses. In defense, they make nucleases, termed *restriction endonucleases* or restriction enzymes, that break DNA at one particular sequence. For instance, the intestinal bacterium *Escherichia coli* makes the enzyme EcoRI that breaks DNA at the sequence -G-A-A-T-T-C-. Different bacteria make different enzymes, recognizing and breaking different sequences of DNA bases. The bacteria label their own DNA at this sequence, adding an extra carbon atom to a few of the bases. The endonuclease does not break this labeled DNA, so the bacterium's own genetic material is safe. But an infecting virus is not similarly protected, so its DNA is broken into harmless pieces.

Restriction endonucleases would remain a mere curiosity but for their immense utility in recombinant DNA technology. Restriction endonucleases are custom molecular scissors, allowing scientists to cut a piece of DNA in any place they please. One merely has to find the species of bacteria that makes the right endonuclease. Hundreds of different enzymes are now commercially available. This ability provided the seed for the entire industry of biotechnology. Today, by recombinant DNA technology, new strains of bacteria are engineered to make commercial quantities of insulin (page 147) or growth hormone (page 144). New strains of plants are engineered, more resistant to frost or disease. And perhaps in the future, biotechnology may allow for the correction of hereditary diseases, such as sickle cell anemia and cystic fibrosis.

Chemical Energy

Once food is digested and absorbed, the work has only begun. Inside our cells, food molecules are broken into their individual atoms and harnessed for energy. Molecules rich in carbon atoms and hydrogen atoms are the best fuels—they are readily combined with oxygen from the air. The "burning" of these atoms is harnessed as chemical energy, locked in *ATP* (see below). Nitrogen, sulfur, and phosphorus atoms are less useful as sources of energy. They are vital, however, as the raw materials for building proteins and nucleic acids. But excess amounts, if any, are merely discarded.

The amount of energy available from the burning of foods is often expressed in Calories (with a capital "C"—scientists often use calories, with a lower-case "c," which are 1,000 times smaller). Calories are measured by actually burning a small sample of a given food and measuring how much the flame heats a measured sample of water. Fats, because they are almost pure carbon and hydrogen, are the richest source of chemical energy, typically providing about 9.5 Calories per gram. Carbohydrates, with their abundant oxygen atoms, provide about half as much energy per gram, and proteins, which are rich in oxygen and nitrogen, provide even less. To perform the basic functions of life—maintaining body temperature, fueling the heart, orchestrating numerous housekeeping chemical reactions—we require about 1,400 to 1,700 Calories per day, depending on body weight. Any additional activity—walking, singing, eating, even thinking—requires additional fuel.

ATP (Adenosine Triphosphate)

In an automobile, gasoline ignites with oxygen, forming an explosion of heat and pressure; metal cylinders capture this energy and convert it into physical work and motion. In an oven, natural gas burns with oxygen, and the heat is used to denature proteins or char carbohydrates, cooking a steak or baking a loaf of bread. We, however, have no metal cylinders or earthen ovens, and cannot withstand this violent release of energy. We cannot burn food in the familiar manner. Instead, we combine food atoms with oxygen one careful step at a time, transferring the morsels of energy to the nucleotide *ATP* (adenosine triphosphate).

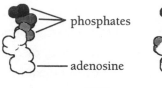

phosphates

adenosine

ATP

ATP is our currency of chemical energy. It captures the energy derived from the burning of food and releases it on demand to other cellular processes. The key to ATP, allowing it to capture and release chemical energy, is its string of three phosphates. Phosphate—a phosphorus atom surrounded by oxygen atoms—carries a strong negative charge. Since like charges repel one another, three individual phosphates normally spring away from one another. But in ATP, they are chemically bonded together, in close proximity. Consequently, it is difficult to make ATP, bringing these three phosphates close together, but easy to break it into pieces, allowing the phosphates to separate. It is so difficult to make ATP that energy from the consumption of food (or, in plants, the energy of sunlight) must be used to connect these three phosphates. And it is so easy to break these bonds that the destruction of ATP is used to move our muscles, heat the body, and power our most difficult chemical reactions.

An average adult stores enough energy to last several months under starvation. Chemical energy is not stored in the form of ATP—ATP serves merely to shuttle energy from supply to demand. Instead, energy is stored in more stable molecules, which may be converted into ATP when needed. Our fluctuating day-by-day requirements are provided by stores of *glycogen*, a large carbohydrate composed of branching chains of glucose. But our normal supplies of glycogen would last only about half a day. Long-term supplies, saved against hard times, are stored in fat. Because fats are very rich in carbon, they form large droplets, with very little water inside. This makes them a compact molecule for storage of energy. Under extreme starvation, after reserves of glycogen and fat are exhausted, more extreme measures are taken. The proteins making up the body are broken down for energy, starting first with the expendable proteins in muscle.

Glycolytic Enzymes

The central pathway of ATP production is *glycolysis* ("sugar breaking"). In 10 successive reactions—breaking a bond here, shuffling an oxygen there—the glycolytic enzymes break glucose into two smaller molecules of *pyruvate*. One particularly energetic step, where hydrogen is removed and placed on the carrier molecule NAD (for more on NAD, see page 72), is used to build ATP. The glycolytic enzymes have different shapes and sizes, suited to their particular function. Some, like phosphoglycerate kinase, are composed of a single protein chain that snaps shut around their target molecules. Others, like glucose phosphate isomerase, are composed of two identical chains, each performing its task separately. In phosphofructokinase, four chains tightly grip one another, working together to regulate the action of the entire enzyme. Together, these 10 enzymes begin the conversion of glucose into usable energy. Nearly all carbohydrates are funneled into this central furnace of energy production, and many other food molecules feed into it at one step or another.

Glycolytic Enzymes

hexokinase

glucose-6-phosphate isomerase

phosphofructokinase

aldolase

triose phosphate isomerase

glyceraldehyde-3-phosphate dehydrogenase

phosphoglycerate kinase

phosphoglycerate mutase

enolase

pyruvate kinase

As glucose is consumed, pyruvate molecules build up and the limited supplies of NAD are saturated with hydrogen. Both the pyruvate and the hydrogen atoms must be used or discarded in some manner. Organisms that breathe oxygen, like ourselves, use these molecules to make even more energy (see below). Other organisms, however, use glycolysis as their only source of energy, and must devise other methods of dealing with the waste products. In the oldest, and the most widely exploited, application of biotechnology, yeast convert their excess pyruvate and hydrogen into alcohol and carbon dioxide. In the making of wine, the sugars in grape juice are fermented by yeast into alcohol. Very dry wines have all of their sugar converted to alcohol, whereas in sweeter wines, the yeast are removed before all the sugar is consumed. Because the yeast need no oxygen, fermentation may be carried out in sealed bottles. The carbon dioxide then builds up inside and forms the festive bubbles of champagne and beer.

Under special circumstances, we can live without oxygen for a short time. Our muscles, in particular, often use oxygen faster than the blood can deliver it, particularly during strenuous "anaerobic" exercise. The ATP used in these spurts of energy, as the levels of oxygen drop, is derived solely from glycolysis. Normally, the pyruvate and hydrogen atoms formed at the end of glycolysis are completely consumed by combination with oxygen. But without oxygen, they build up. So our muscles temporarily combine the two, forming lactic acid and incurring an *oxygen debt*. The rapid buildup of lactic acid leads to muscle fatigue and the pain of overexertion. After the exercise, lactic acid is transported to the liver by way of the blood, where it is converted back to pyruvate. There, pyruvate may be used to build new glucose, which is delivered back to the muscles, or it may be combined with oxygen in the normal way to form ATP. Some bacteria use this same mechanism to ferment sugar, using the ATP for energy and discarding the lactic acid. The bacteria in sourdough bread use glycolysis to break the starches in flour, releasing the lactic acid into the dough, giving it its pungent sour taste.

Citric Acid Cycle

Since we breathe oxygen, we have access to a powerful process for energy production, capable of making 18 times more ATP per glucose molecule than that produced through simple glycolysis. Instead of discarding the atoms in pyruvate, as yeast do, we burn them with oxygen and use the energy released to build ATP. The burning of pyruvate begins in an enormous enzyme factory, the *pyruvate dehydrogenase complex*. In three steps, it strips a carbon atom from pyruvate and releases it as carbon dioxide. This geodesic complex is an enzymatic assembly line, performing the three chemical reactions in rapid progression. The many protruding arms act as conveyor belts, moving each pyruvate from one enzyme to the next.

The remaining piece is broken into its component atoms in the *citric acid cycle*. Its hydrogen atoms are stripped away and placed on NAD, for use in the final stage of energy production, *electron transport* (see below). Its two carbon atoms are converted into carbon dioxide and delivered to the blood, to be exhaled. As with the glycolytic

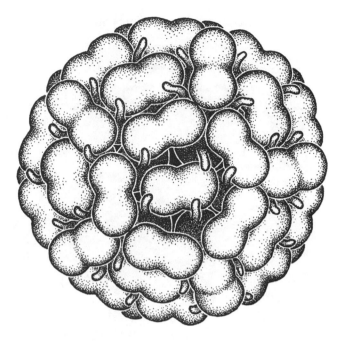

Pyruvate Dehydrogenase Complex

enzymes, the enzymes of the citric acid cycle are diverse, each suited to their individual chemical task. Two are particularly unusual. The fourth, α-ketoglutarate dehydrogenase, is a large, multienzyme complex similar to the complex that acts on pyruvate. It also acts as a molecular factory, shuttling molecules through several steps with its many flexible robot arms. The sixth, succinate dehydrogenase, does not float freely inside cells. Instead, it is bound in the membrane of the mitochondria. In its reaction, it extracts hydrogen atoms from the slowly dwindling molecule and passes them directly to the energy-producing machinery of the mitochondria.

The "cycle" refers to the cyclic nature of the carrier that chaperones the pyruvate atoms through the process. First, they are attached to oxaloacetate to form citric acid (citric acid, not to be confused with ascorbic acid, or vitamin C, is commonly found in citrus fruits, lending the pleasing sweet/sour taste to lemons and oranges; it is also added to sour confections). The remaining steps of the cycle progressively strip off the hydrogen atoms and add oxygen atoms to the carbon atoms, forming carbon dioxide. At the end, oxaloacetate is restored, ready for the next cycle.

Citric Acid Cycle Enzymes

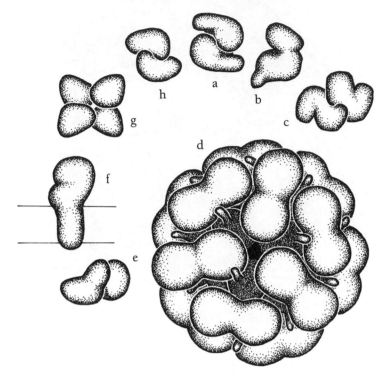

a. citrate synthase
b. aconitase
c. isocitrate dehydrogenase
d. a-ketoglutarate dehydrogenase complex
e. succinyl-CoA synthetase
f. succinate dehydrogenase
g. fumarase
h. malate dehydrogenase

Electron Transport Proteins

During glycolysis and the citric acid cycle, the hydrogen atoms in food molecules are removed and placed on the carrier molecule NAD, along with an extra electron, forming NADH (NAD, nicotinamide adenine dinucleotide, is built from ATP and the vitamin *niacin*. Niacin contains a particularly reactive ring of atoms, which safely carry hydrogen atoms and electrons and ATP provides a familiar handle for the enzymes using NAD). Most of the energy that heats, moves, and controls the

Electron Transport Proteins

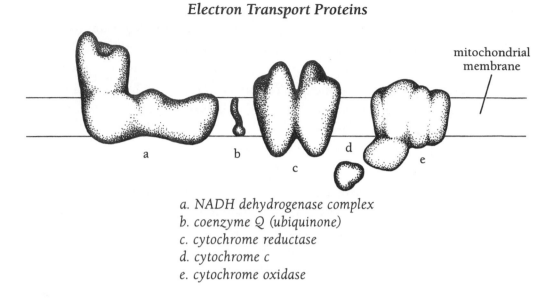

mitochondrial
membrane

a. NADH dehydrogenase complex
b. coenzyme Q (ubiquinone)
c. cytochrome reductase
d. cytochrome c
e. cytochrome oxidase

body is produced from this NADH, after the food molecules themselves have been completely broken into their component atoms. The two electrons carried by NADH—one in the hydrogen atom and one extra—are used to make ATP in a truly surprising manner: they are consumed to charge a cellular battery, which then fuels the production of ATP. The batteries are the mitochondria, large compartments scattered through each of our cells. As described with ribosomes (page 49), mitochondria are thought to be the descendants of individual bacterial cells that took up residence in animal cells billions of years ago. Today, mitochondria are dedicated factories, specializing in the conversion of hydrogen atoms into usable ATP energy.

Four proteins and one small lipidlike molecule together charge the battery. The process begins with the *NADH dehydrogenase complex*, a large protein complex that reclines in the mitochondrial membrane, reminiscent of a Henry Moore sculpture. It removes the electrons from NADH (releasing a hydrogen ion into the surrounding water in the process) and channels them down the *electron transport chain*. The electrons flow through the NADH dehydrogenase complex, then hop to *coenzyme Q*. Coenzyme Q floats through the membrane and passes the electrons to the third, *cytochrome reductase*, a large protein complex immersed in the membrane. Again, the electrons hop through the protein, ultimately being added to *cytochrome c*. Cytochrome c is a small protein that is freely soluble in water. It bumps through the surrounding solution, passing the electrons to the final protein complex, *cytochrome oxidase*. *Cytochrome oxidase* finally places the electrons on oxygen gas, and hydrogen ions from the surrounding water (released earlier) are added to the electron-rich oxygen atoms, forming water.

As electrons flow through these five molecules, like electricity in a wire, the current is used to pump hydrogen ions across the membrane. Hydrogen ions pile up outside the mitochondria, charging the battery. The pressure of these overcrowded hydrogen ions fuels the production of ATP. The hydrogen ions are allowed to flow one by one back into the mitochondria, but not for free. They return through the enzyme *ATP synthase*, powering the synthesis of ATP (below) as they go. Nearly all of our energy is generated by the charging of this battery, so this pathway is very sensitive to poisons. A poison that attacks any of the proteins in this chain will kill in minutes, as scant supplies of ATP are rapidly exhausted. Cyanide is the classic example. It binds to an iron atom in cytochrome oxidase, blocking its action. The electron transfer chain is blocked and each cell asphyxiates, bathed in oxygen but unable to use it.

ATP Synthase

ATP synthase is a molecular waterwheel, harnessing a swift flow of hydrogen ions to build ATP. It sits immersed in the mitochondrial membrane, supporting a round head on the inner side. Hydrogen ions flow through ATP synthase back into the mitochondrion, powering the rotation of the round head relative to the membrane-bound base. Each revolution requires the energy of about nine hydrogen ions returning into the mitochondrion. This mechanical motion is converted into chemical work. The head contains three identical active sites, each of which builds an ATP molecule with every turn. In a fully charged mitochondrion, the flow of hydrogen ions turns ATP synthase at a speed of 100 to 200 revolutions per second, manufacturing three ATP molecules with each turn.

ATP synthase fuels "aerobic" exercise. Slow and steady, aerobic exercise allows enough time for the blood to deliver adequate oxygen to our muscles. Anaerobic exercise, on the other hand, is fast and frenetic, working our muscles faster than can be supported by the blood. Anaerobic exercise must be fueled by more primitive means, as described in with glycolytic enzymes (page 68). Because of the greatly increased amounts of

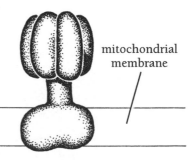

mitochondrial membrane

ATP Synthase

ATP made by complete breakdown of sugar, aerobic exercise may be sustained for much longer than anaerobic exercise. Track events are a case in point. Short sprint events rely almost entirely on anaerobic energy. Sprinters reach incredible speeds, but only for short times, as their muscles wastefully ferment their energy reserves. Long events, on the other hand, are entirely aerobic, testing endurance more than peak performance. Imagine a marathon runner with heart and muscle cells filled with fields of ATP synthase molecules rotating at rapid speed, churning out a constant supply of ATP.

Hemoglobin and Myoglobin

Oxygen is needed in every corner of the body for making ATP aerobically. One might envision building tiny air passages throughout the body, allowing oxygen to flow directly to each cell. This is unfeasible for a body of our size, but is used in the smaller bodies of insects. Instead, we use blood to shuttle oxygen from the air to our cells. Unfortunately, oxygen is not particularly soluble in water, so merely circulating water is not enough. Instead, our blood is filled with hemoglobin, an oxygen-carrying protein. Hemoglobin grabs oxygen where it is plentiful, as the blood passes through the lungs, and releases it where it is scarce, in the muscles and tissues. The hemoglobin in a liter of blood can carry about the same number of oxygen molecules as are found in a similar volume of air: blood is liquid air.

Hemoglobin is perhaps our most visible protein. Its four iron atoms provide the deep red color to blood. Hemoglobin does not float freely in the blood; instead it is packed into red blood cells, which shuttle it around the body. Red blood cells are dedicated completely to their job, so much so that they cannot be called "living" any longer. As red blood cells develop, they fill themselves to capacity with hemoglobin. At the last moment, they gather up all of their cellular machinery—DNA, enzymes, RNA, ribosomes—and eject them, leaving only a simple shell surrounding a sea of hemoglobin molecules. They are then placed in the blood, where they circulate and carry oxygen for about four months. Red blood cells have been reduced to the role of supertankers: a huge hold, filled to capacity with hemoglobin. Their sole task is the transport of oxygen from the lungs to the tissues of the body.

Hemoglobin is composed of four protein chains, each tightly holding an iron atom inside a disk-shaped heme group. Hemoglobin is an allosteric protein, using a change in shape to enhance its function much like the enzyme aspartate

75

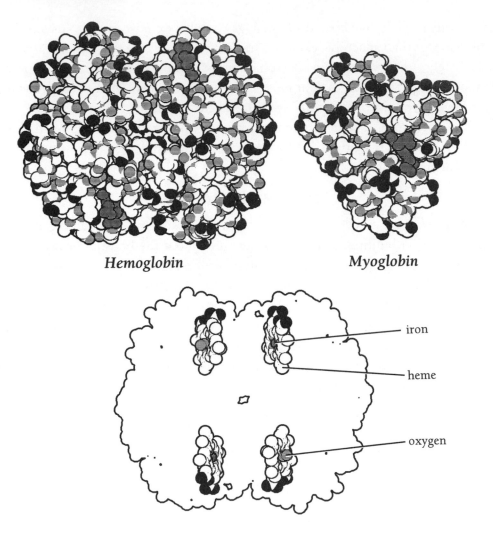

Hemoglobin   *Myoglobin*

iron

heme

oxygen

carbamoyltransferase (page 23). When oxygen is plentiful, one iron atom may easily capture an oxygen molecule. When it does, it forces the surrounding protein into a slightly different shape. This shape pulls on the surrounding three chains, making them more receptive, and these three sites quickly capture oxygen molecules. When oxygen is scarce, these same allosteric motions have the opposite effect. One of the four protein chains may lose its oxygen and revert to its original shape. It then pushes on the remaining chains, making them less comfortable with their oxygen molecules. Hemoglobin is perfectly designed for oxy-

76

gen transport: in the lungs, hemoglobin soaks up oxygen like a sponge; in the tissues, hemoglobin wrings out the last traces of oxygen for our hungry cells.

Myoglobin is a simpler version of hemoglobin, composed of only a single protein chain, with a single iron–heme complex. Each myoglobin molecule holds only one oxygen atom, making it less efficient at picking up and dropping oxygen at will. Myoglobin is used as a storehouse for oxygen, particularly in the muscles. The muscles of whales are deep red because of their rich supplies of myoglobin, to store the oxygen needed for extended dives. The myoglobin molecule shown here is the molecule from sperm whale muscle.

Ferritin and Transferrin

We each have just under 4 grams of iron in our bodies—a little less than the weight of a nickel. Most is used for oxygen transport: 2.5 grams is locked in hemoglobin and another 0.1 to 0.2 grams in myoglobin. A small additional amount is distributed in other enzymes, such as those in the electron transport chain (page 72). The rest is safely stored inside our cells in *ferritin* containers. Ferritin is composed of 24 protein subunits arranged into a hollow shell. About 3,000 iron atoms can be stored in each ferritin complex, along with neutralizing ions of hydroxide and phosphate.

Free iron ions rapidly form insoluble compounds in water, becoming completely unusable. Iron must be continually protected from water. We transport iron ions inside protective *transferrins*, along with neutralizing carbonate ions. Ferritin specializes in the storage of large amounts of iron, transferrin in the delivery of indi-

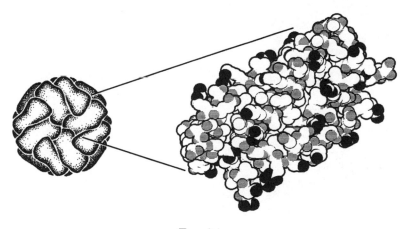

Ferritin

vidual atoms. Transferrins circulate in the blood, and the similar *lactoferrin* is found in milk and other fluids. Cells pluck transferrins out of the blood, extract their iron ions, and release the empty proteins back into the blood, to scavenge for more. Because these proteins bind so tightly to iron, they also are efficient bactericides, killing bacteria by depriving them of their necessary iron. Therefore, transferrins are also found in egg whites, where they provide iron to the growing embryo and also help protect it from infection.

Iron is jealously guarded because of its essential role in oxygen transport. It is efficiently recycled when red blood cells wear out: we lose less than 1 milligram of iron each day. Careful recycling is necessary because iron is difficult to absorb. Metallic iron is useless—only iron *ions* can be utilized. In the stomach, the iron in protein foods is released by acid. Vitamin C then reduces it to the ionic form which is more easily absorbed. This process is inefficient, however, and only one in 10 iron ions is actually absorbed.

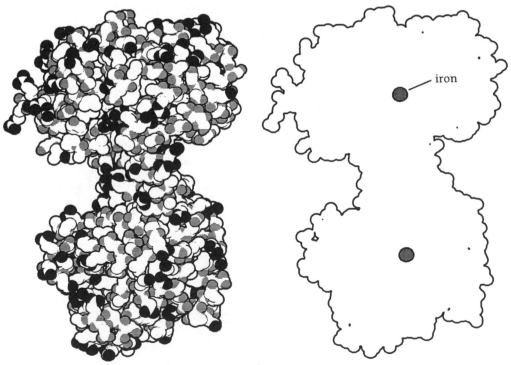

Transferrin

Adenylate Kinase

Some enzymes, such as those that digest fats, harness the energy of ATP by breaking the inner of the two phosphate–phosphate bonds, releasing a chain of two phosphates called *pyrophosphate*. Breakage at this site instead of the normal outer site provides slightly more chemical energy for these specialized reactions. Unfortunately, our energetic machinery—the enzymes of glycolysis, the citric acid cycle, electron transport, and ATP synthase—is designed to replace a single phosphate, not to add two new phosphates. *Adenylate kinase* is the solution to this problem.

With the help of an atom of magnesium, it transfers a phosphate from ATP to AMP (adenosine monophosphate, with one phosphate), forming two molecules of ADP (adenosine diphosphate, with two phosphates). The two ADP molecules then enter into the normal energy-producing processes. An additional enzyme—*inorganic pyrophosphatase*—breaks the pyrophosphate in half, forming two free phosphates that may also enter into the reactions.

active site

Adenylate Kinase

Adenylate kinase is a compact enzyme with a deep active site groove. Its reaction is sensitive to the presence of water: if a water molecule got between the ATP and the AMP, the phosphate might be attached to it instead, breaking ATP into ADP and a free phosphate, and leaving AMP untouched. This would be a complete waste of resources. The deep, buried active site of adenylate kinase, however, closes around AMP and ATP when they bind, completely shielding the reaction from water. Many enzymes that use ATP have this same property—an active site that shelters the molecules being transformed.

Nucleoside Diphosphate Kinase

Our energy-producing machinery is designed for the production of ATP: quickly, efficiently, and in large quantity. The dozens of enzymes described in this chapter digest everything we eat and use the pieces, ultimately, to turn ATP synthase. But we also need other nucleotides besides ATP. Thymine, guanine, uracil, and cytosine nucleotides, as well as adenine, are needed to build the various forms of RNA, and to build DNA when the cell divides.

Nucleoside diphosphate kinase is the bank where chemical currency is exchanged. It transfers the phosphate from a nucleoside *tri*phosphate, such as ATP, to a nucleoside *di*phosphate molecule, such as guanosine diphosphate (GDP). The reaction is performed through a "ping-pong" mechanism. The first nucleoside triphosphate binds in one of the six active sites, and the phosphate is grabbed—ping!—by a histidine (the enzyme shown here has three active sites visible on the front surface, and another three on the back). Then, a different nucleoside diphosphate binds to the same site, and the phosphate is attached to it—pong!—to form a new triphosphate.

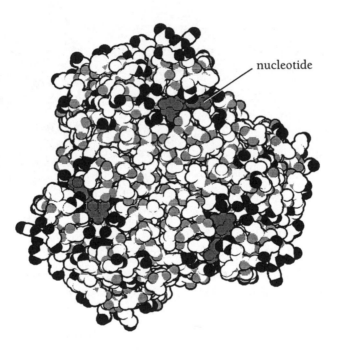

nucleotide

Nucleoside Diphosphate Kinase

4
Form and Motion

Cells are inventive architects. Using only invisibly small building blocks, cells fabricate tough fibers of protein, providing flexible strength to tendons and cartilage. By embedding mineral crystals in these fibers, cells build bones and teeth—stony structures strong enough to last millions of years. By filling themselves to bursting with vanishingly thin ropes of protein, skin cells form a flexible, insulating coat. Linking these ropes tightly together provides additional strength for fingernails. A dense scaffold of protein supports and directs the convoluted inner world of each cell. To build these elaborate structures, some thousands of times larger than an individual cell, one can find examples of any engineering principle in use today. Fences are built, railways are laid, reservoirs are filled, and houses are constructed complete with rooms, doors, windows, and even decorated in attractive colors. Lap joints, buttresses, waterproofing, reinforcing rods, valves, concrete, adhesive—each has a molecular counterpart.

The molecules of form and motion are our most familiar molecules. We interact with these molecules every day. Strands of keratin add silky resilience to our hair. If we are unhappy with their shape, we may physically cut these molecules shorter, or chemically straighten or curl them. Melanin colors our skin from palest tan to deep brown. By lying in the sun, we may induce our skin cells to make more, darkening the color. Legions of molecular motors, toiling along highways of protein, contract our muscles as we walk and breathe. Through repeated exercise, we can persuade our muscle cells to make more motors, increasing our strength. In this chapter, we explore these most personal molecules, beginning with the molecules that shape and support each cell, continuing with the molecules that shape the entire body, and finishing with the molecules that power motion.

Cell Form

Our cells are tiny, water-filled soap bubbles. Just as soap bubbles trap a pocket of air inside a flexible membrane of detergent, cells corral their molecular machinery with a flexible membrane of *lipids*, shielding them from the external world. Of course, a perfectly seamless skin would not do; cells must be able to interact with the outside. A perfectly sealed cell would wither from lack of food or fester in an accumulation of wastes. To deliver food inside and release wastes outside, cells build a battery of protein doorways through their cell membranes. Spanning the membrane, these proteins supervise the constant traffic of molecules into and out of cells.

Like a soap bubble, this outer skin is fragile, so our cells build a *cytoskeleton*, microscopic bones and muscles, to provide support to their tenuous skins. Just as we are supported by a system of rigid bones, our cells are supported by a system of protein filaments. The cytoskeleton, however, is not a permanent structure like the skeleton. It is continually assembled, destroyed, and reassembled according to need as the cell stretches, divides, shrinks, or crawls. The long, sturdy filaments are built of small proteins, stacked like bricks in a chimney. These small building blocks make the cytoskeleton easy to build and quick to disassemble.

Lipid Bilayer

A thin skin of lipid encloses each of our cells, forming our outermost barrier to the surrounding world. Walls of lipid also provide privacy to different areas inside our cells, allowing mitochondria to concentrate on energy production, the nucleus to focus on information handling, and lysosomes to focus on destruction of obsolete molecules. Lipids form large cellular structures—sheets and spheres and tubes—in a manner different from the cellular structures we have seen thus far. Proteins, nucleic acids, and carbohydrates all form large structures by chemi-

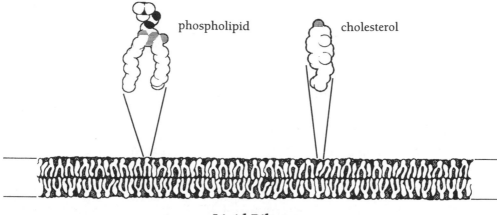

Lipid Bilayer

cally linking many small building blocks in a long chain. Lipids, on the other hand, form large structures by *stacking* many small molecules together, like bricks in a wall, instead of permanently linking them together. Lipids are composed of two parts: a charged head group, which is freely soluble in water, and one or several carbon-rich tails, which are highly insoluble in water. When placed in water, lipid molecules cluster together to shelter their tails inside, away from the water, but at the same time attempt to leave their head groups exposed to the water. Many shapes are possible: the lipids may form a small, round ball, with tails inside, called a *micelle*, or a large, flat sheet, with head groups facing outward on either side, called a *lipid bilayer*. We use small micelles to transport lipids and bilayers to form the outer skin of each cell.

Because lipid bilayers are composed of so many small, independent molecules, they are highly fluid. Individual lipids slide past one another, rotate in place, and even flip-flop between the inner and the outer faces. In spite of this dynamic behavior, lipid bilayers are persistent barriers, resisting the passage of molecules into and out of cells. Ions such as sodium and potassium are strongly excluded, because their strong electric charges abhor the oily interior of the membrane. Ions must be deliberately transported into and out of cells, by proteins such as the calcium pump (page 88) and the sodium–potassium pump (page 159). Proteins and nucleic acids also do not cross membranes, simply because they are too large to penetrate the smooth lipid coat. Even smaller molecules like amino acids and sugars cannot pass from one side to the other, without assistance. Only very small, uncharged molecules dissolve in the membrane, passing into and out of our cells at will. Oxygen seeps through the walls of capillaries, feeding the surrounding cells. Tiny molecules of alcohol rapidly permeate our cells, intoxicating every corner of the body.

Phospholipids are the most common lipids in our cell membranes. Their head groups contain a phosphate and a small molecule such as serine, the sugar inositol, choline, or ethanolamine. The phosphate carries a negative charge and the small molecule is typically rich in oxygen and nitrogen atoms, making the head groups of phospholipids very soluble in water. A molecule of glycerol connects this head group to two tails, varying in length from about 12 to 16 carbon atoms. These tails carry names descriptive of where they were first discovered, such as palmitic acid and oleic acid. The tails may be maximally covered with hydrogen atoms, making them straight and flexible. These "saturated" chains pack densely next to one another, forming thick, sluggish membranes. Or pairs of hydrogen atoms may be removed, forming stiff bends in the chain. These "unsaturated" chains loosen up the packing and improve the fluidity of membranes.

Cholesterol is also common in our cell membranes. It is a rigid lipid composed almost entirely of carbon and hydrogen, making it virtually insoluble in water. Its angular corners, like the kinks in unsaturated lipids, make a membrane more fluid. Because cholesterol is rich in carbon, membranes containing cholesterol are sealed more tightly than membranes composed only of phospholipids. In particularly flexible cells, cholesterol may comprise up to half of the total lipid. Red blood cell membranes are rich in cholesterol: they must be highly flexible to allow them to squeeze through the narrow capillaries that feed our extremities. Cholesterol also provides the raw material for several important hormones. The steroid sex hormones (page 149) circulated during puberty are made from cholesterol. Sunlight falling on our skin converts another cholesterol into vitamin D, which is necessary for proper maintenance of calcium in the bones. Cholesterol has a darker side as well. Excess cholesterol is normally excreted with bile from the gallbladder. Trouble with this process can lead to gallstones. About three-fourths of gallstones are crystalline aggregates of cholesterol, the remainder are stones of calcium salts. Cholesterol is also the key player in atherosclerosis, the formation of hard deposits inside arteries.

A menagerie of proteins interacts with lipid bilayer membranes to assist in the transport of molecules and messages across the barrier. There are several ways in which a large protein can interact with a sheer sheet of lipid. Perhaps the simplest method is to attach a single lipid to the protein. The lipid then inserts into the membrane like a normal lipid, tethering the protein to the membrane like a dirigible tied to a mooring tower. This approach is taken by the G proteins (see page 153), which are attached firmly to the inside of the cell membrane. A second method is to build a protein with a lipid surrogate: a long tail of valine, leucine, and other carbon-rich amino acids. This tail may then insert into the membrane, perhaps crossing it entirely, firmly anchoring the protein to the surface. Aminopeptidases (page 60), growth hormone receptors (page 152), and imunoglobulin E (IgE) antibodies (page 124) use this approach. The third method is far more extreme: a protein is built with a band of carbon-rich amino acids encircling its entire form. These proteins insert bodily into the membrane, aligning this band with the interior of the membrane like an inner tube encircling a bather. Porin (see below) is an example. Because lipid membranes are so fluid, all these proteins—tethered, anchored, or immersed—are free to move in the membrane unless they are attached to something inside the cell. They rapidly rotate and often bump into one another as they travel through the confines of their two-dimensional world.

Porin

The simplest method to allow molecules into a cell is to punch a hole through the cell membrane, allowing free passage. Of course, this is not a good idea in most cases: a leaky cell membrane would allow nutrients to escape and toxins to creep in. But in special cases, a leaky membrane is just what is needed. Our mitochondria are surrounded by two concentric membranes. The inner membrane holds all of the protein machinery for generating energy, including the electron transport proteins (page 72) and ATP synthase (page 74). Because it acts as a cellular battery, storing a charge of hydrogen ions on one side, the inner membrane must be sealed for proper function. The outer membrane, however, is needed only to keep proteins such as cytochrome c (page 72) close to the inner membrane, so that they do not wander off to other parts of the cell. This outer membrane must be leaky to allow entry of fuel. Mitochondrial porin forms perfect cylindrical holes of just the proper size to allow pyruvate and NADH in but not to allow proteins out. Porin is composed of three identical subunits, each forming a watery tunnel through the normally impervious membrane.

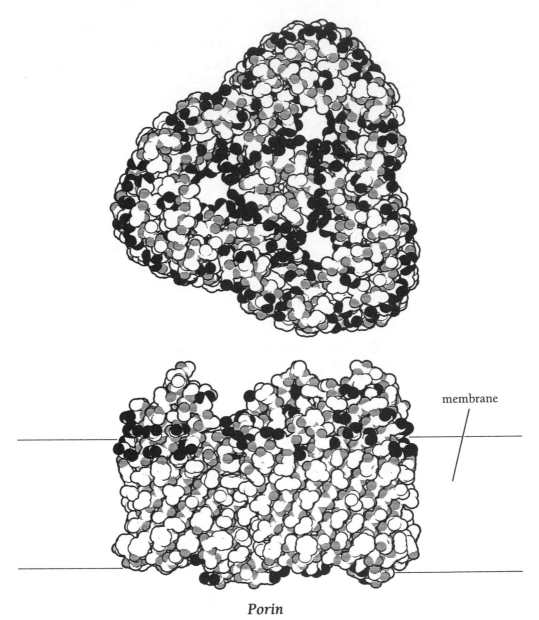

Porin

Many bacteria are similarly composed of two concentric membranes, an outer leaky one and an inner sealed one. In between are proteins that sort and classify the molecules entering the cell, and proteins involved in energy production similar to those in the mitochondria. The porin shown here is a bacterial form, found in the bacterial outer membrane. Notice, in the side view, the band of carbon-rich amino acids (predominantly white) encircling the molecule, making it comfortable inside a membrane. The surface left exposed at top and bottom, and through the tunnel, seen in the top view, is studded with charged amino acids (predominantly gray and black), allowing them to interact comfortably with the water on either side of the membrane.

Connexons

The human body is a friendly community of trillions of cells. None live in isolation; all communicate and cooperate for the good of the community. For long-distance communication—from organ to organ and from brain to limbs—cells send messages to one another through the blood and through the nervous system (see Chapter 5). But for neighboring cells, such as those in the layered tracts making up the skin, communication is direct. They simply talk to their neighbors over the backyard fence, through the tiny holes in connexons.

Cells are sewn together, almost touching, by thousands of connexons at small patches called *gap junctions*. Each connexon is composed of six identical protein units, together forming a hexagonal tube through the cell membrane. A constant traffic of sugars, amino acids, ATP, and other small molecules travels from cell to cell through them. However, large molecules like proteins cannot pass through these narrow tubes, so each cell retains its own machinery. In times of distress, these tiny knotholes can be sealed. The concentration of calcium inside cells is normally kept very low, so if large amounts of calcium flow through a

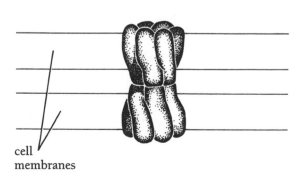

cell
membranes

Connexons

cell, it is usually a signal that it has been breached. Sensing a sudden rise in calcium, connexons snap shut, isolating healthy cells from a damaged neighbor.

Nuclear Pore

Inside the nucleus, at the center of each cell, is the information needed to build each of our proteins; outside in the cytoplasm is the machinery that actually builds them. The nucleus serves to protect the precious strands of DNA, in which all of the hereditary information is stored. It is surrounded by a double layer of membrane, separating the bustle of the cell from the quiet archives of the nucleus. The nuclear pore is the port of communication between the nucleus and the rest of the cell—the front door of the library.

Nuclear pores are some of the largest and most complex protein assemblies that we build. A pore is composed of a ring of eight identical units, each composed of dozens of different proteins. Small molecules such as nucleotides and sugars travel freely through nuclear pores. Larger molecules, however, must be guided. Proteins whose jobs lie in the nucleus, such as nucleosomes (page 39) and polymerases (page 40), are made in the cytoplasm and must be transported into the nucleus. These proteins are built with an extra piece at the end, which acts as a signal sequence. Nuclear pores search for proteins with these *signal sequences* and pull them inside. Conversely, messenger RNA (page 42), transfer RNA (page 47) and the RNA in ribosomes (page 49) are all made inside the nucleus, and must be transported out to perform their cytoplasmic duties. These RNA molecules are chaperoned outside by proteins, which again use a signal sequence as their ticket for transportation.

Calcium Pump

Calcium is our most abundant mineral, but most of our kilogram and a half of calcium is locked away inside the bones. Only a few grams circulate in our tissues, but this small amount performs many diverse roles. Calcium provides chemical leverage to some enzymes, such as starch-digesting amylase (page 61). Calcium strengthens the interactions of the blood-clotting machinery (page 134), guaranteeing a sure-footed response to damage. Calcium is also used as a messenger inside cells. We typically lose about half a gram of calcium each day, and require a replenishing source. If the amounts available in the diet, from milk and green vegetables are not sufficient, calcium may be extracted from the vast supplies stored in the bones.

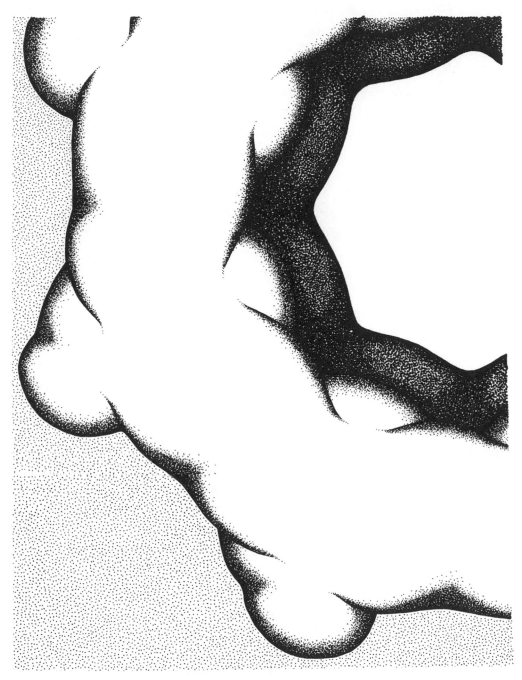

Nuclear Pore

Cells normally deplete themselves of calcium: the normal level of calcium inside each cell is about 10,000 times lower than the level outside. Because calcium levels are so low, calcium is a rare sight inside cells. A handful of calcium ions spreading through a cell are easily recognized, delivering a quiet, but insistent, message. The curiously humanoid calcium pump, embedded in the cell membrane, maintains calcium at this vanishingly low concentration, and removes excess calcium after a message is sent, leaving the cell clean for the next. Powered by ATP, it pumps calcium ions one at a time from inside to out.

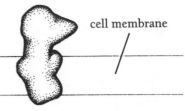

Calcium Pump

The smooth contraction of muscles is controlled by tides of calcium. Receiving a message from the brain, muscle cells release calcium from an internal storage area, and the wave of calcium stimulates them into contraction. Calcium pumps then quickly replace it back into storage, ready for the next action. The entire process is amazingly fast: the ebb and flow of calcium, in less than a second, controls the rhythmic systole and diastole of every heartbeat.

Actin

The cytoskeleton supporting each cell—bracing the fragile cell membrane and organizing all of the internal compartments—is built of bricks of actin. Like bricks, actin is stacked to build enormous, sturdy structures. Also like bricks, a sharp force can break obsolete structures, to be rebuilt later in new form. Actin molecules stack one atop the next to form a helical filament. A network of crisscrossing actin filaments fills nearly every cell, providing a scaffold on which our enzymes perform their functions. Actin filaments allow each cell to change its shape according to its individual duties. White blood cells scavenge through the blood by pushing out arms and feet from their cell surface with filaments of actin, crawling like an amoeba. Digestive cells use bundles of actin to push brushlike fingers into the intestine, providing more surface for the absorption of nutrients. Muscle cells use ranks of parallel actin filaments as a ladder on which myosin (page 105) climbs.

Phalloidin, a poison from the death cap mushroom, causes uncontrolled growth of actin filaments, with deadly result. Coma and death result after about a week as phalloidin and the related poison α-amanitin (see RNA polymerase, page

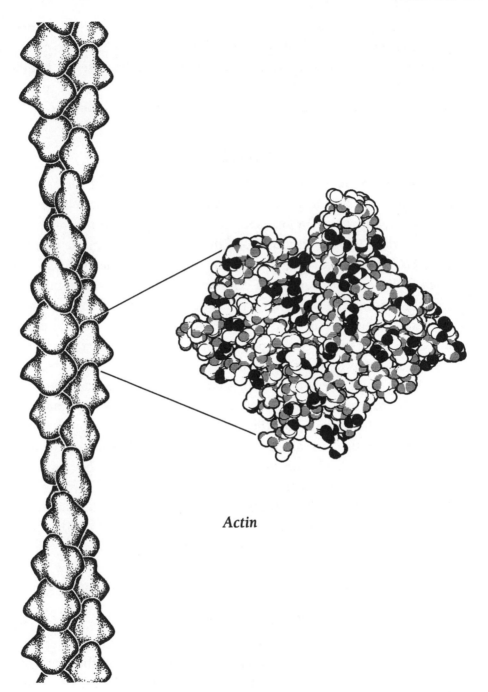

Actin

44) attack cells in the liver and kidneys. Milk is often given as an antidote for poison because milk fats and milk proteins neutralize many poisonous molecules. But for phalloidin a large quantity of meat works better, because the bundles of actin in muscle fibers soak up the poison. Of course, for this antidote to be effective, the mistake of eating this poisonous mushroom must be discovered before symptoms occur.

Microtubules

Microtubules are the railways of our cells. They are the largest filaments, composed of protein bricks arranged to form a hollow chimney. Cargo is transported along microtubules by protein locomotives. This transport is directional: *kinesins* drag their burdens in one direction along the tubule, *dyneins* in the other direction (page 107). In a typical cell, microtubules fan out from the nucleus and wrap around the cell periphery, forming a cellular railway system. Microtubules also perform other specialized tasks. In nerve fibers, they supply strength to the long, narrow axons and mediate the transport of resources along their enormous lengths. Microtubules extend the length of sperm tails, and when slid next to one another by dynein motors, power their swimming.

When a cell divides, microtubules form the railways used to pull the two daughter cells apart, separating one set of chromosomes into each. Some poisons and drugs act by interfering with this vital function of microtubules. The drug *taxol*, isolated from the bark of the yew tree, causes uncontrolled growth of microtubules. When treated with taxol, cells lose control of the microtubules needed to separate their two sets of DNA and get stuck midway through the process of division. Ultimately, the drugged cells die. Because of this property, taxol is widely used as an anticancer drug. Cancer cells divide more rapidly than normal cells, and thus are more sensitive to attack by taxol. Unfortunately, taxol also attacks other normally dividing cells, such as those in hair follicles and those lining the stomach, causing the severe side effects of cancer chemotherapy.

Intermediate Filaments

Unlike actin filaments and microtubules, intermediate filaments are very stable: once built, they resist disassembly. The reason for this stability lies in the shape of their component proteins. Instead of stacked proteins, as in actin filaments and microtubules, intermediate filaments are built of interlocking units. The long, dogbone-shaped

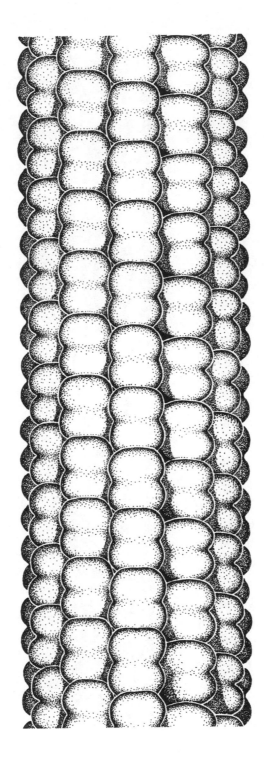

Microtubule

proteins firmly grasp their neighbors to form knobby filaments.
Intermediate filaments form a
tough network throughout our
cells, often woven crosswise to
actin filaments to strengthen the
tangled cytoskeleton. Intermediate
filaments may also be used alone
to build a sturdy structure. A
dense sheet of intermediate filaments strengthens the inside of
the nucleus, with multiple layers,
each running perpendicular to the
next.

Intermediate filaments provide flexibility to skin and toughness to fingernails. We are each
covered with a layer of dead cells
strengthened with *keratin*, an intermediate filament protein. As
they grow, skin cells fill themselves
to bursting with keratin. At the last
moment, they activate enzymes to
crosslink the keratin, which may
comprise 85% of the total protein
at that point, and promptly die.
These scaly dead cells form a
protective layer, shielding us from
sunlight, damage, and disease.
Similarly, dead cells filled with
keratin are extruded into hairs or
molded into fingernails.

Different keratins are
built for different needs. All have
the characteristic knobby shape of

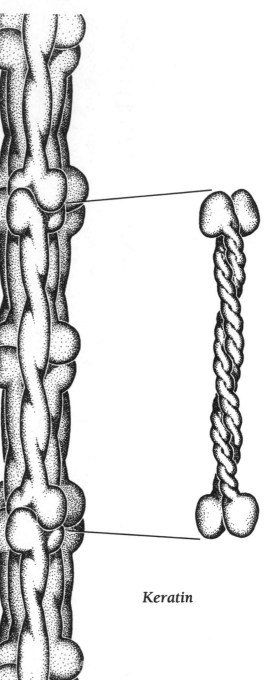

Keratin

intermediate filaments, but each differs in local details. Our skin keratins contain many sites where glutamate and lysine may crosslink, linking one protein to a neighbor. More extensive crosslinking, performed just before the cell dies, changes pliable skin into tough calluses. Our hair keratins are rich in cysteine. Hair derives its strength from the additional crosslinks formed between the sulfur atoms in these amino acids. Fingernails use both: mixing one-fifth skin keratins with four-fifths of the tougher hair keratins. The crosslinks in hair may be artificially broken and reformed in a permanent wave. The sulfur bonds are broken, making the hair pliable. After designing the desired curl, the crosslinks are reformed. Because of the sulfur chemistry involved, the process can be quite odorous.

Body Form

The textures of the body arise from the structural molecules layered between our cells, and not, as one might expect, from the textures of the cells themselves. Rubbery fibers laced just under our skin cells add flexibility and elasticity. Molecular cables form the wiry tensile strength of tendons. Molecular concrete, reinforced by rigid molecular rods, forms the stony matrix of bones and teeth. These molecules are built for strength. The structures inside cells, described in the previous section, are most often composed of small globular pieces, stacked like bricks; they are easily built and quickly disassembled. Structures outside cells, on the other hand, are large and built to last. They are extensively overlapped and chemically crosslinked to one another. When we have finished with one of these structures, we cannot merely disassemble it and recycle the parts. We must digest away the area and begin anew.

Collagen

Collagen is our most plentiful protein, comprising about one-fourth of our total protein. Belying its ubiquity, it is perhaps the simplest protein. It is composed of three similar protein chains wrapped tightly around one another in a triple helix, forming a long rod. A repeated sequence of three amino acids forms this unusual structure. Every third amino acid must be a tiny glycine, because every third amino acid is packed tightly in the interior of the helical cable. In addition, one of the two amino acids between these repeated glycines is usually proline or hydroxyproline, forming kinks that direct the chain back into the helical rod at each turn. Hydroxyproline is made by adding an oxygen atom to proline, in a reaction that requires vitamin C. Scurvy is caused by a deficiency of vitamin C, which slows the synthesis of hydroxyproline and ultimately stops the construction of new collagen. The signs of scurvy—the loss of teeth and easily bruised skin—are due to the lack of collagen to make repairs on the normal wear and tear of eating and working.

We make several types of collagen, each with a long piece of triple helix attached to a different functional end. The simplest is merely a long, almost featureless triple helix with blunt ends. Molecules of *type I collagen* associate side by side in staggered registration to form a tough fibril, often called a "banded fibril" because the gaps between individual molecules are easily seen in the electron microscope. Banded collagen fibrils crisscross the space between nearly every one of our cells. *Type IV collagen*, on the other hand, has a small globular element at one end. This increases its repertory of structural interactions. The globular domains associate head to head, tying two molecules tightly together. At the opposite end, four molecules bind together, two facing in one direction and two in the other to form an X-shaped structure. Finally, the molecules associate side by side in various registrations. Together, many type IV collagen molecules lock together to form an extended meshwork, used to form tough structural layers, such as the basement membrane that separates the outer layer of skin from the more delicate inner layers. *Type VII collagen* has an unusual three-armed structure attached to its triple helix. These collagen molecules associate tail to tail, placing bundles of these three-

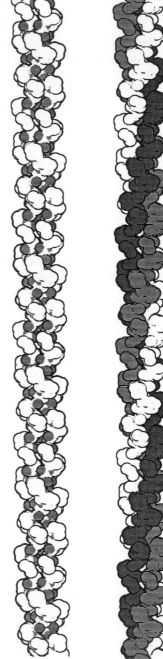

Collagen

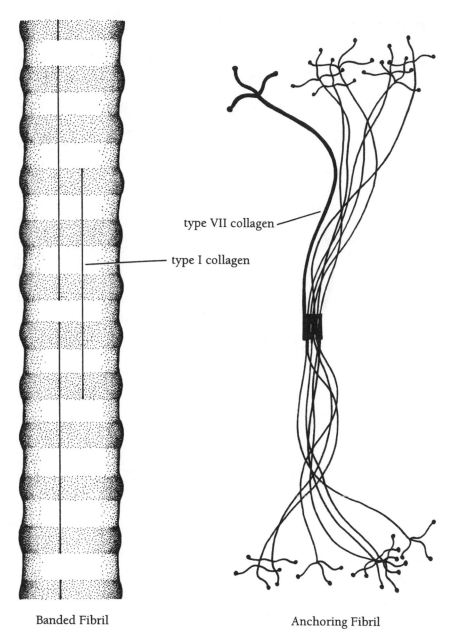

type VII collagen

type I collagen

Banded Fibril   Anchoring Fibril

(200,000X)

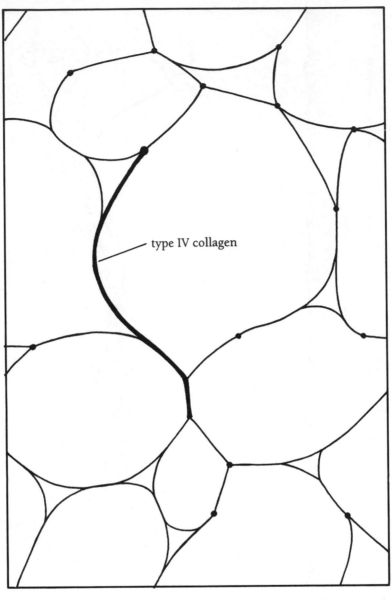

type IV collagen

(200,000X)

armed structures at opposite ends. Bundles of type VII collagen sew banded fibrils to basement membranes and are therefore known as "anchoring fibrils." (Note that the illustrations of these three collagens are at a lower magnification than the other illustrations in this book because of their large size.)

For some structural tasks, we need a material even stronger than dried keratin fibers, stronger than collagen fibers, stronger than hair, skin, and cartilage. For these tasks, we grow tiny, perfect mineral crystals. In the bones, cells build a typical matrix of collagen around themselves, but then pack crystals of apatite, a mineral of calcium and phosphate, between and around the fibers. The result is a material harder than nails, but not quite as hard as stone. Bone cells entomb themselves in mineral and protein, leaving only threadlike holes leading to the blood, through which they siphon nutrients. Many living things grow mineral crystals to strengthen their bodies. Mollusks build sea shells out of calcium carbonate, in the form of aragonite and calcite. Diatoms build glittering glass houses out of silica, like the silica in sand. Some bacteria grow tiny crystals of ferrite, a magnetic iron mineral, and use them to distinguish north from south in their muddy ponds. It is thought that migratory birds may also have this ability, perhaps due to similar magnetic crystals.

Boiling the collagen from livestock unwinds the triple helix, forming tangled chains commonly known as *gelatin*. Gelatin "gels" when cooled because the tangled chains soak up large quantities of water, like a sponge. Commercial gelatin is produced from cartilage and tendons, which owe their toughness to rich tracts of collagen.

Glycosaminoglycans

Concrete alone is not sufficient for building bridges and buildings: it is brittle and prone to cracks. Iron rods alone are not sufficient either: they are malleable and prone to bending. But when iron reinforcing rods are embedded in concrete, a tough structure is formed, resistant to sharp forces and capable of carrying heavy loads. Our connective tissue, layered between cells, uses the same principle: strong cables of collagen reinforce a concrete matrix of glycosaminoglycan.

Glycosaminoglycans are composed of tangled masses of carbohydrates. Each chain of sugars traps a layer of water along its entire length; together, many tangled chains form a solid, gelatinous mass of carbohydrate and water. The same principle applies when starch is used to thicken a sauce. Starch normally occurs in tight granules, but boiling extends the long chains and they grab the surrounding water, thickening the gravy. *Hyaluronic acid* is the simplest glycosaminoglycan, composed of a single unbranched chain of carbohydrate thousands of sugars long. The proteoglycans, such as *chondroitin sulfate, dermatan sulfate, heparin,* and *keratan sulfate,* are more complex,

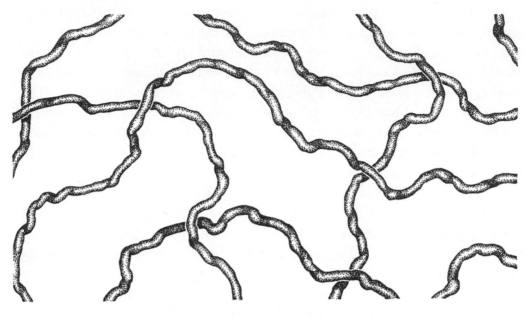

Hyaluronic acid

formed of many shorter carbohydrate chains tethered to a long central protein chain. We are glued together and supported by these tough gels, reinforced with collagen.

Apart from providing mechanical support to the spaces between cells, glycosaminoglycan gels can also act as molecular filters. The gel is somewhat porous, allowing passage of smaller molecules but holding back larger molecules. The kidneys use just such a gel to filter waste products from the blood. As blood flows through the kidney, urea easily passes through the gel and is ultimately concentrated into urine. The important proteins in blood, such as fibrin (page 135), and antibodies (page 124), are too large to fit between the tangled strands of carbohydrate and are retained in the blood vessels.

Elastin

Elastic proteins similar to rubber provide flexibility and resilience to our tissues, allowing us to bend without breaking. Vulcanized rubber is composed of long, flexible chains of carbon, arranged in tangled coils and connected to one another by linkages of sulfur. When rubber is stretched, these chains unwind and extend but are held relative to one another by the linkages. Without the linkages, the

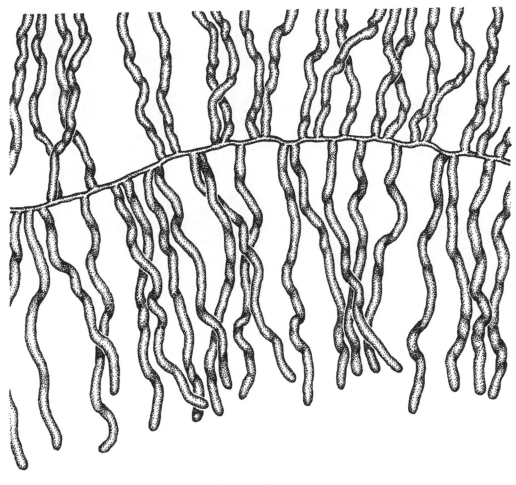

Proteoglycan

chains would slide and the rubber would tear. But with the linkages, neighboring strands stay neighbors and the rubber stretches. When the force is released, the chains spring back to their original, relaxed coils, and the rubber returns to its original shape.

Elastin uses the same principle. Elastin is unusually rich in the small, flexible amino acids glycine and alanine, and contains many proline kinks, ensuring that the elastin chains form a random tangle instead of the normal compact protein shape. At about 40 places along its length, each elastin chain is linked to three neighboring chains by an unusual linkage: four lysine residues chemically

101

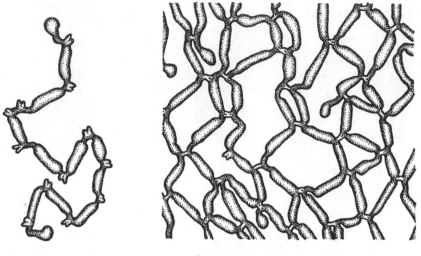

Elastin

react, clasping hands to form a rigid pack-saddle. The linked elastin molecules form an elastic tangle with properties similar to rubber.

Elastin is found throughout the body. Our vocal cords are composed primarily of elastic tissue, and a layer of elastic fibers allows our arteries to beat with the constant rhythm of high and low blood pressure. The cartilage in our ears is rich in elastic fibers. Only two other natural elastic molecules are found in the animal kingdom: *resilin* is found in the hinges of insect wings, and *abductin* is found in the hinges of mollusk shells. Many plants, such as rubber trees, make elastic *latex*, which dries to form a rubbery coating over wounds in the bark.

Fibronectin

Fibronectin and the related molecules *tenascin* and *laminin* act as molecular adhesives. They are large proteins with several flexible arms. Arrayed along each arm is a string of sticky sites. Some sites adhere to cell surfaces, and other sites affix to collagen and glycosaminoglycans. These molecules also adhere to one another, the two arms of fibronectin forming long cables and the branched arms of laminin forming loose, felty layers. These sticky molecules act as flexible adapters, gluing cells to the surrounding connective tissue.

Fibronectin, in several similar forms, is found throughout the body. Most is used to adhere cells to the collagen and glycosaminoglycans of connective tis-

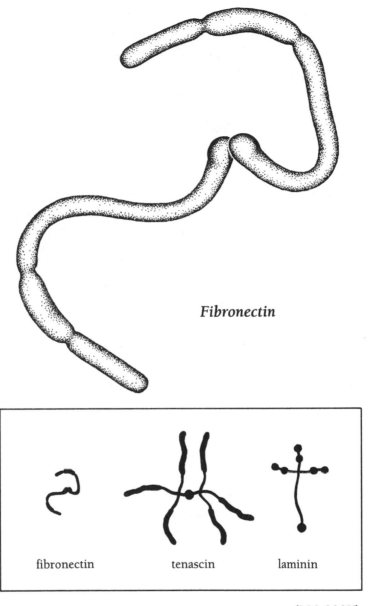

Fibronectin

| fibronectin | tenascin | laminin |

(200,000X)

sue, the nails and staples that hold us together. Another form circulates in the blood, where it aids in recognition of a wound, triggering the formation of a blood clot (see Chapter 5). Fibronectin is also essential in the development of embryos. Embryonic cells crawl along highways of fibronectin, folding layers, pulling long fibers, forming pockets and knobs, as limbs, heart, and brain are modeled from formless tissues. This embryonic form of fibronectin also finds use in adult life. As a wound is healing, fibronectin provides the pathways over which skin cells grow and migrate to fill the damaged area.

Melanin and Tyrosinase

Melanin, of all of our thousands of different molecules, has been the source of the most human strife. Melanin is the pigment that provides the attractive range of colors to human skin, from palest pink to deepest brown. It is sequestered in tiny granules dispersed though our skin cells, where it serves to absorb and scatter dangerous ultraviolet radiation. Different skin colors result from different sizes and different numbers of melanin granules, mixed with the pale pink of blood in capillaries.

We make two forms of melanin: *eumelanin*, which is densely brown, and *phaeomelanin*, which is reddish yellow. Brown and black hair is colored primarily by eumelanin, and phaeomelanin is the principal pigment in red hair. Blond hairs contain very little pigment, nor do hairs that have gone gray with age. Eye color is determined by the melanin content of the irises. Brown eyes are richly colored with melanin. Blue eyes, on the other hand, contain no melanin. The unpigmented irises scatter blue light more efficiently than other colors, so they appear blue, for much the same reason that the sky is blue. Shades of hazel and green are due to graded mixtures of scattered blue light with earth tones from small amounts of melanin.

Melanin is made from tyrosine by the copper-containing enzyme *tyrosinase*. It converts tyrosine into a reactive intermediate, which then spontaneously links into large, random aggregates of melanin. Because of its random nature, different parts of the molecule absorb different colors of light, together absorbing all colors and looking black. Exposure to ultraviolet light causes skin to make additional melanin, tanning in the process. Tanning occurs in two steps. *Immediate tanning* peaks one or two hours after exposure to light and fades within one day. It is a short-term response, as the skin uses the tyrosine it has readily available. *Delayed tanning* occurs after a more extensive exposure to ultraviolet radiation. The skin makes a more significant investment in tyrosine and melanin production, peaking about a week after exposure and producing coloration that lasts for months.

Molecular Motors

Life is motion: plants gracefully follow the sun, and animals frenetically chase after food and attempt to avoid becoming food. Plants typically use hydraulics to power their motion, filling and emptying bags of water to bend stems and lift leaves. Our most familiar motions—flexing arms and legs, beating heart, and breathing lungs—are powered by muscle cells, which derive their power from tiny molecular motors. Spending the energy held in a molecule of ATP, each tiny motor bends, contracting the cell by one atom-sized step.

Motors also power most of our other, less familiar motions. Molecular motors typically bridge two sets of rigid structures, moving one relative to the other. White blood cells crawl through the blood, looking for invaders, by forcibly shifting their cytoskeletons. Cells lining the respiratory tract wave tiny oars, sweeping along the protective layer of mucus, by sliding microtubules next to one another. Sperm cells also move by sliding microtubules, forcing their tails into serpentine waves.

Myosin

Myosin provides the power for running and walking, for lifting and holding, for all of our voluntary motions. Myosin also powers involuntary muscles, the beating heart and the slow squeeze of digestion. Muscle cells are filled with cables of actin (page 90), alternating with thicker filaments composed of the motor protein myosin. Together, actin and myosin form the engine of muscle contraction. Myosin may comprise more than half of our muscle protein, and another quarter is actin. The rest of the proteins serve to fuel and control the engine.

Myosin is composed of two large heads connected to a long, thin tail. About 400 myosin molecules combine by aligning their tails, staggered one relative to the next. The head groups extend from the surface of the resulting myosin filament like the bristles in a bottle brush. The bristling head groups provide the power to contract muscles. They reach from the myosin filament to a neighboring actin filament and attach to it. Breakage of an ATP molecule then forces the head into a radically different shape. It bends near the center and drags the myosin filament along the actin filament. This is the *power stroke* of muscle contraction. In a rapidly contracting muscle, each myosin head may stroke five times a second, each stroke moving the filament about 10 nanometers (about the width of an

Motor Proteins

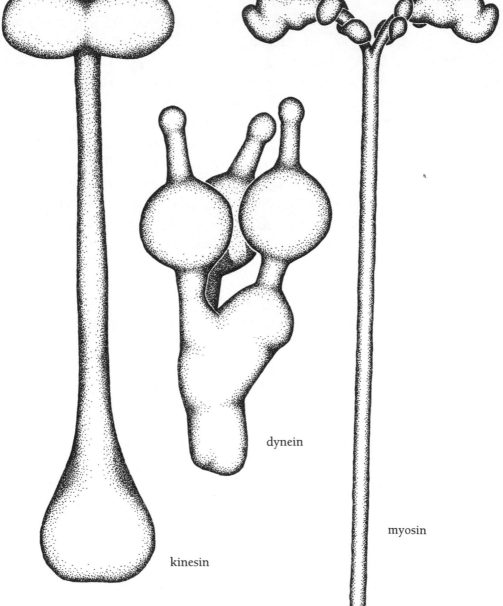

kinesin

dynein

myosin

average protein). Countless infinitesimal myosin power strokes together yield the muscle strength we each experience.

Strength training and body building increase the bulk of muscles by adding new tracts of actin and myosin to each muscle cell, increasing their size. Thicker, stronger connective tissue is also built to strengthen connections, and new blood capillaries are constructed to deliver additional oxygen and nutrients. Examples of hysterical strength—a mother lifting a car off her child—are thought to be due to the simultaneous contraction of all of the necessary muscle cells, a feat that is impossible to perform voluntarily.

Kinesin and Dynein

Inside our cells, molecules are transported along rails of microtubules (page 92). Two proteins perform the work: *kinesin* drags objects in one direction, *dynein* in the other. Both use ATP as their source of chemical energy, converting it into physical motion. Kinesin is reminiscent of myosin. It has two globular heads, which perform a myosinlike power stroke, and a long tail, which attaches to the object to be moved. Dynein, on the other hand, is entirely different, composed of two or three large head groups, which perform the motion, connected to a large base, which attaches to the cargo.

Dynein powers the swimming of sperm and many of the more subtle motions of cells. The tail of a sperm cell contains a long bundle of microtubules, with dynein bridging from one to the next. By forcing these filaments to slide relative to one another, dynein bends the tail into the sinusoidal curves used to propel the sperm. Our lung cells build tiny cilia that extend into the air passages and beat by a similar mechanism. These tiny oars move a layer of mucus that coats and protects the delicate internal membranes.

5
Dangers and Defenses

We are under continuous assault. The elements erode at our every surface. Toxic metal salts poison susceptible proteins and pollutants destroy others. Water and oxygen slowly corrode our molecules. The world is filled with physical dangers: spines, corners, abrasives, and sharp edges. And on top of these dangers, microorganisms relentlessly probe for toeholds wherever we are exposed to the exterior world.

The skin is relatively impervious to these attacks, forming our first line of defense. But the delicate membranes of the lungs, which must be exposed to absorb oxygen, and the delicate surfaces of the digestive system, which must remain exposed to absorb nutrients, remain easy targets. In defense, we build a diverse system of protective molecules to fight this continuous onslaught. Dedicated proteins fight the most common foes, such as hyperreactive forms of oxygen and heavy metals. But for more insidious enemies, the immune system wages a targeted battle, customized to fight each new foe.

Toxins and Venoms

A few million molecules of botulism toxin, far less than the weight of a single grain of salt, are enough to kill an adult man or woman. The human body, in all of its interconnected complexity, can be quite fragile. Each molecule depends on every other. If one key mechanism is disabled, the entire system falls. Poisons typically attack the three most essential, and thus vulnerable, bodily systems. The quickest poisons, such as curare and strychnine, target the nervous system, causing instant paralysis, often leading to death by asphyxiation. Poisons that block cellular respiration and energy production, like cyanide and carbon monoxide, also kill in minutes, as the body rapidly uses up its meager reserves of oxygen and ATP. The slowest poisons, such as α-amanitin from poisonous mushrooms, attack protein synthesis. These poisons act slowly, over days. Death often results from liver failure, as the liver is the major site of protein synthesis.

Toxic *enzymes* are particularly effective in their dark task. Simple chemical poisons such as carbon monoxide attack proteins one-on-one: each carbon monoxide molecule binds to one iron atom in hemoglobin (page 75), blocking its action. Toxic enzymes, on the other hand, go from molecule to molecule, destroying one after the next. In some cases, such as diphtheria toxin, a single enzyme molecule can kill an entire cell.

Snake Venom

Poisonous snakes inject a pharmacopoeia of unpleasant substances in their strike. Deadly nerve and heart toxins are mixed with enzymes that break down tissues and cells, allowing the venom to penetrate further into the unfortunate victim. These include lipases that break cell walls and collagenases that digest collagen (page 95) in connective tissues. In developed countries, most snake bites are not fatal, because effective antitoxins are available to neutralize the poisonous proteins in snake venom. But the digestive enzymes in venom often destroy the tissue surrounding the bite, so medical treatment may be needed as the bite heals over the course of several months.

Cobras and coral snakes inject a deadly neurotoxin: a protein that attacks nerve cells. It binds to the acetylcholine receptor (page 163), blocking signals from nerve to nerve and from nerve to muscle. The muscles never receive their signals to contract, causing paralysis. A cardiotoxin is also present, which destroys the charge across cell membranes that is normally used to propagate a nervous signal. The heart is its primary target, since the heart requires nerve signals to pace every beat. Both toxins are very small proteins that insinuate themselves into the nerve machinery, acting as a molecular "monkey wrench" lodged in the system.

Animal Venoms

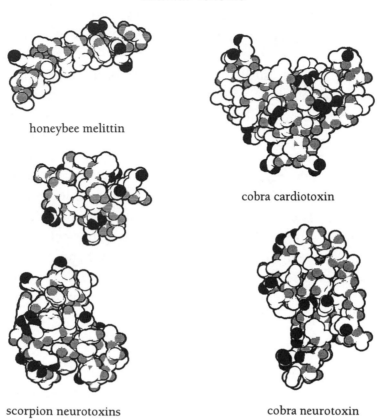

honeybee melittin

cobra cardiotoxin

scorpion neurotoxins

cobra neurotoxin

Antidotes for snake venoms are composed of antibodies (page 124) that neutralize the deadly toxin proteins. They are produced by injecting minute amounts of venom into domestic animals and isolating the antibodies that they produce. (Only a small amount of venom may be injected at any time, yielding only a small amount of antibody, for the obvious reason.) These antibodies may then be injected around the site of a snake bite, neutralizing the venom in place.

Spider Venom

Spiders are arguably the most feared of the venomous animals, perhaps because they are found everywhere. Barely a day goes by without coming across one creeping somewhere inside the house or outdoors. Most are venomous, but only a handful carry a venom strong enough or in large enough quantities to pose us a real threat. Black widow spiders are the best known; perhaps you have stumbled over the beautiful glossy body of a female black widow, and with a little trepidation, entrapped one in a glass jar to see the bright red hourglass. Black widows are shy and usually require this type of insulting behavior to induce them to bite a human. A neurotoxin in the spider venom is the primary toxic element. Surprisingly, a black widow bite may be painless. But severe symptoms soon follow, as the toxin progressively attacks the nervous system. Within an hour, the area of the bite reddens and the marks of the fangs may become apparent. As the venom spreads, red lines extend from the bite, along lymph ducts in the area. The lymph nodes swell and become painful, and the pain spreads to the muscles. Tremors may occur, often causing a painful grimace of the face. At the same time, the toxin is affecting the nerves involved in thought, resulting in extreme feelings of dread, consuming restlessness, and bouts of crying. These severe symptoms last for several days, but are lethal in only one in 20 bites. Recovery is slow—several months of weakness and pains—as the body rids itself of the venom.

Scorpion Venom

Scorpion stings are a major problem in tropical climates. The retiring habits of scorpions—resting just under the surface of sandy areas, thus easily stepped on, or hiding in shoes and clothing after their nocturnal foraging—often bring them into contact with people, particularly in cultures without secure housing. Fortunately, scorpion stings, although very painful, are rarely lethal. The major toxic components of most scorpion stings are nerve toxins. Each species of scorpion typically makes several similar toxins. They are small proteins with about 30 to 70 amino acids, depending on the particular toxin and the particular species. The small protein chain is interconnected by three or four linkages between cysteines, making these toxins stable and resistant to destruction. After a sting, a progressive destruction of nervous communication becomes apparent. Soon after a person is stung, the most sensitive nerves show problems: mental function is impaired, the victim is often nervous and excited, and sight may be blurred. More serious consequences follow as the toxin spreads to more robust systems: irregular pulse and breathing, changes in body temperature. In a few cases—one in 20 with the most dangerous species—death follows in about one day, due to cardiac arrest or respiratory failure.

Bee and Wasp Venom

Scorpions, spiders, and snakes use their venom to immobilize prey. They most often use a nerve toxin to paralyze their victims instantly. Bees and wasps, however, make their venom for protection. Their venom is designed to be acutely painful, to act as a sharp warning instead of a lethal blow. The pain of a bee or wasp sting is due primarily to histamine (page 142). Histamine normally signals an inflammatory response, but the quantities in a bee sting overload the system. Wasps and hornets also inject excessive quantities of neurotransmitters such as serotonin and acetylcholine (page 161), which short-circuit local nerve and muscle control. Bee venom also contains high levels of *melittin*, a tiny protein that breaks down blood cells. Of the 26 amino acids in melittin, five are charged—enough to make melittin quite soluble in water (and blood). The remaining portions are carbon-rich amino acids like leucine and valine, making melittin equally happy in the carbon-rich environment of a cell membrane. It is thought that melittin attacks cells by pounding tiny wedges into the membrane, forming cracks and rifts that ultimately breach the cell.

These toxins are not sufficient in themselves to be lethal. However, an allergic reaction to the proteins in bee and wasp stings can be life-threatening. Soon after a bee sting, antibodies (page 124) may be made to neutralize some of the proteins. These antibodies remain in the blood, attached to white blood cells, after the toxin is cleaned up. Normally, these cells control an inflammatory response. When they sense this toxin again, they release compounds like histamine, signaling to the surrounding area that something is wrong. The local blood vessels dilate, flooding the area with blood and allowing lymph from the surrounding tissue to enter. But in cases of bee sting allergy, this response is amplified to lethal proportions. After a second sting, these cells release their histamine so rapidly that large tracts of blood vessels simultaneously dilate, causing the blood pressure throughout the body to drop sharply. This can send the victim into a lethal *anaphylactic shock*.

Bacterial Toxins

A single molecule of the toxin made by diphtheria bacteria can kill an entire cell. Botulism and tetanus toxins are millions of times more toxic than chemical poisons like cyanide. These bacterial toxins are designed for deadliness—they are the most toxic substances known. They combine a specific targeting mechanism, allowing the toxins to seek out and find susceptible cells, with the toxicity only possible with an enzyme. Once inside an unfortunate cell, the toxin jumps from molecule to molecule, destroying one after the next until the cell is killed.

Cholera bacteria release a toxin that attacks intestinal cells. Cholera toxin is composed of a ring of five proteins, which is the targeting mechanism, and a lollipop-shaped protein, which is the toxic enzyme. The ring seeks out carbohydrates on the surface of intestinal cells and firmly attaches to them. Upon binding, the toxic enzyme is injected into the cell, were it wreaks its havoc. Inside, it searches for G proteins, which normally relay hormonal messages inside cells (see page 151), and chemically connects an ADP molecule to them. This places the G protein in a permanently active state, sending a never-ending hormonal signal to the cell. Normally, trace amounts of hormone signal important events, like hunger. But this overloaded signal confuses the cell, making it transport water and sodium ions uncontrollably outward, flooding the intestine. Severe diarrhea results, sometimes so severe as to lead to death by dehydration. *Enterotoxin*, a similar toxin made by the normal intestinal bacterium *Escherichia coli*, causes a less severe reaction: encountering new strains of the bacterium on a holiday may lead to the discomfort of traveler's diarrhea.

Diphtheria bacteria make an exquisitely toxic enzyme, composed of three functional parts, arranged at the points of a stretched triangle. One part performs the task of binding to the surface of the target cell. Once attached, the toxin lies in wait until it is swallowed by the cell. Inside the cell, acid is normally added to digest food particles. However, acid activates diphtheria toxin instead of digesting it, and the second part of the molecule forces the toxin through the membrane into the cell. Once inside, the third part of diphtheria toxin, which acts as a toxic enzyme, disables every copy of a key protein involved in protein synthesis. Unable to build any new proteins, the cell dies.

The vaccine against diphtheria toxin now makes the disease relatively impotent, but diphtheria was once a major cause of death, especially among children. The vaccine is part of the *DTP vaccine*, which immunizes against diphtheria, tetanus, and pertussis (whooping cough). The diphtheria and tetanus portions are composed of actual diphtheria toxin and tetanus toxin that have been inactivated with formaldehyde to render them harmless (the pertussis vaccine is composed of entire killed bacteria). These crippled versions of the toxins, termed

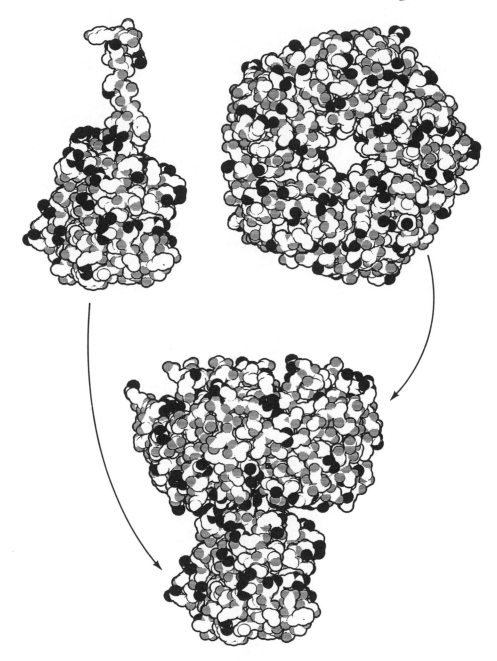

Enterotoxin

toxoids, fool the immune system into making antibodies (page 124), which will later destroy any real toxin released by an infection.

Tetanus bacteria and botulism bacteria are both sensitive to air, being killed by oxygen, and only attack humans under special conditions. Botulism is typically encountered in food poisoning, when the bacteria grow in improperly canned foods. When the cans are opened, the bacteria die as they are exposed to oxygen, but the toxin in the food remains just as effective as when they were alive, sealed away from the air. Tetanus is encountered in infections of stab wounds, perhaps from stepping on a rusty nail. The bacterial spores are embedded deep within tissues, away from the air.

Both tetanus and botulism toxins are composed of two parts, as in diphtheria toxin and cholera toxin. One part binds to the surface of the cell and the other enters and attacks the cell. They are effective neurotoxins, acting at the junction between nerve cells and muscle cells, but with markedly different effects. Botulism toxin blocks excitatory nerve junctions, blocking the signal for the muscle to contract. Severe botulism poisoning results in paralysis, with all muscles completely relaxed, unable to contract. Tetanus toxin, on the other hand, blocks inhibitory nerve junctions, blocking the signal to release contraction. Tetany results, with uncontrolled spasms of muscles. The characteristic clenching of the jaw muscles leads to the common name *lockjaw*.

Detoxification

The environment is filled with toxic compounds. Some are formed by the action of ultraviolet light on normal molecules, such as corrosive ozone formed from oxygen. Some are made by the plants and animals that we eat, such as bitter tannins and alkaloids. And some have a modern genesis in the organic chemistry laboratory, including industrial chemicals and preservatives. These reactive substances attack our delicate molecular machines, crippling or inactivating them. They must be destroyed, or at least made less dangerous. We make a collection of detoxification enzymes expressly for this purpose. They are tailored to protect against common enemies: attackers that will almost invariably be seen during our lifetime. For more exotic attackers, we rely on a more powerful system—the immune system—which is discussed in the next section.

Cytochrome p450

A family of several dozen enzymes, collectively termed cytochrome p450 enzymes because of their brilliant color, are built to connect oxygen atoms to carbon-rich molecules. These enzymes are remarkable for their lack of specificity, in contrast to the precise reactions performed by most enzymes. Using an iron atom held in a heme group similar to that in hemoglobin (page 75), cytochrome p450 enzymes will take any of hundreds of different molecules—chemical solvents, antibiotics, hallucinogens, pollutants, vitamins—and add oxygen atoms to them. These varied molecules are similar in that all are relatively insoluble in water. By

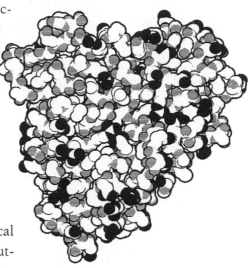

Cytochrome p450

adding oxygen atoms, they are made more soluble and thus easier to flush out of the body. Apart from this role in detoxification, p450 cytochromes also perform many synthetic reactions, when an oxygen atom is needed. Vitamins A and E are converted into active forms, and cholesterol is converted into steroid sex hormones (page 149) in part by these enzymes.

Cytochrome p450 enzymes are most common in the liver. When food is digested and absorbed, food molecules are delivered through the blood directly to the liver, where the molecules are processed and sorted. The liver has the first chance to absorb and store nutrients, for instance, storing sugars in the form of glycogen. It also carries the task of intercepting poisonous substances, clearing them away before they can reach more susceptible tissues. Cytochrome p450, bound to liver cell membranes, converts these molecules into harmless derivatives for removal by the kidneys. As one might expect, the liver suffers the major damage in cases of poisoning. Alcoholics continually poison the liver with alcohol—if their liver cells die faster than they can be regenerated, structural cells take over, clogging the liver with cords of connective tissue, causing *cirrhosis*.

Detoxification by cytochrome p450 can be a double-edged sword. Drugs are often destroyed by cytochrome p450, reducing their useful lifetime. Headache remedies, such as acetaminophen, must be taken every few hours because cytochrome p450 is continuously raking them out of the blood. The sensitivity of individuals to different drugs is often a consequence of the efficiency of their collection of cytochrome p450 enzymes. In rare cases, the products of the reaction can be even worse: the oxygenated form can be even more toxic than the original. Many carcinogens, such as those in cigarette smoke, are converted into active forms by p450 cytochromes.

Glutathione

Glutathione plays several essential roles in our protection. It is composed of three amino acids connected in tandem: glycine, cysteine, and in an unusual bond to its acidic group, glutamate. Glutathione is not built by the normal machinery of protein synthesis. The unusual bond to glutamate poses an impossible problem to a ribosome. Instead, it is constructed from its three component amino acids by two custom enzymes. The central cysteine is the key to the protection afforded by glutathione. Its sulfur atom scavenges destructive molecules like peroxides and free radicals, converting them to harmless compounds. In the liver, the enzyme *glutathione S-transferase* takes the sulfur from glutathione and attaches it to toxic molecules, making them more soluble and easier to eliminate, similarly to the

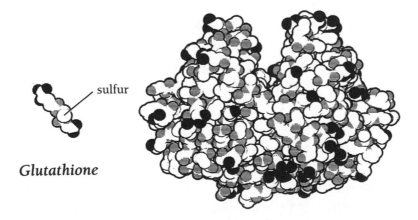

sulfur

Glutathione

Glutathione S-Transferase

oxygen added by cytochrome p450 enzymes (above). Glutathione also maintains our proteins in their proper form. Its sulfur atom reacts with unnatural sulfur–sulfur bonds in proteins, breaking them and allowing the proper pairings to form. This reactive sulfur also maintains the iron in hemoglobin (page 75) at the proper charge. Relatively high concentrations of glutathione, found throughout the body, are necessary for these functions. Cysteine by itself would probably serve as well, but free cysteine is quite reactive and would be toxic at this high level.

Glutathione plays a role in the detoxification of *acetaminophen*, a nonaspirin pain killer. Acetaminophen is broken down first by reaction with a cytochrome p450 enzyme, forming a highly toxic intermediate, then by addition of glutathione, forming a nontoxic product that is promptly excreted. In normal analgesic amounts, the drug is harmlessly cleared away in a few hours. However, if an overdose of acetaminophen is taken—perhaps 30 grams for a normal adult—the reserves of glutathione in the liver are depleted in this reaction. The highly reactive intermediates formed by cytochrome p450 then build up and react with other vital cellular components, causing extensive liver damage.

Superoxide Dismutase

We are continually bathed in a dangerously reactive but absolutely essential molecule: oxygen gas. Most of our energy is generated when oxygen is converted into water by the addition of electrons and hydrogen ions (see Chapter 2). Occasionally, an oxygen molecule will escape before the conversion is complete, resulting in a *superoxide* radical (an oxygen molecule with an extra electron). Superoxide is

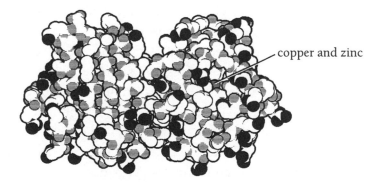

copper and zinc

Superoxide Dismutase

extremely reactive and promptly attacks any surrounding molecules, crippling them. The enzyme *superoxide dismutase* protects us from this danger, capturing superoxide molecules and detoxifying them, using an atom of zinc and an atom of copper. The reaction begins with two identical superoxide molecules. The extra electron on one is removed and transferred to the other. The one losing an electron becomes oxygen gas, the one gaining an electron then grabs two hydrogen ions to form hydrogen peroxide. This unequal reaction of two identical molecules, like one twin brother becoming rich off the misfortune of the other, is termed *dismutation*.

Antioxidants such as vitamin E and vitamin C play a similar role in detoxification. They capture free radicals and render them harmless. Free radicals carry unusually reactive electrons, making them particularly dangerous. When they attack a protein, they form a new free radical in the process. A chain reaction occurs, with the free radical hopping from protein to protein, destroying each in turn. Antioxidants act as suicide molecules, capturing the free radical and rendering it harmless, but sacrificing themselves in the process.

Catalase

We detoxify many alcohols and acids by bleaching them with hydrogen peroxide. Tiny sealed compartments in each cell, termed *peroxisomes*, are filled with concentrated hydrogen peroxide. It is formed as a waste product by enzymes that perform *oxygenation* reactions, removing hydrogen atoms from molecules and replacing them with oxygen atoms. Some of the fats in our diet are broken down in peroxisomes, forming hydrogen peroxide. Catalase then uses this reactive hydrogen peroxide to add oxygen atoms to poisonous molecules, such as small alcohols, beginning the process of elimination. Catalase is composed of four identical protein subunits. The active sites are buried deep within each protein, connected to the surface by a narrow tunnel. An iron ion, held in a heme group similar to that in hemoglobin (page 75), gives catalase the strength to capture reactive peroxide molecules. A quarter of the alcohol from alcoholic beverages is broken down this way—the rest funnels through alcohol dehydrogenase (page 33). Catalase may be so plentiful in these tiny compartments that it forms tiny, perfect crystals, easily seen in the electron microscope.

Outside the peroxisomes, catalase is found in smaller amounts, with a different task. Instead of using hydrogen peroxide, it destroys it. Catalase performs a dismutation reaction similar to that performed by superoxide dismutase, taking two hydrogen peroxide molecules and dismuting them to water and oxy-

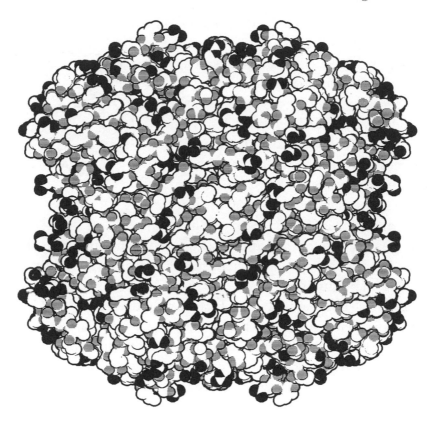

Catalase

gen gas. Hydrogen peroxide is a dangerous molecule to have roaming free in the cell—just think of the chemical change it makes in hair.

Lysozyme

Lysozyme protects us from the ever-present danger of bacterial infection. It was discovered during a deliberate search for medical antibiotics. Alexander Fleming was using the *shotgun method*, still in wide use today. He added everything he could think of to growing bacterial cultures, looking for anything that would slow their growth. He discovered lysozyme in an unlikely source: when he had a cold, he added a drop of mucus to the culture and, surprisingly, it killed the bacteria. Later, lysozyme present in mucus was shown to be the active principle. Unfortunately, the size of lysozyme makes it unusable as a drug. It would work as a topical drug, killing bacteria in the

121

areas to which it was applied, but it could not rid
the entire body of disease, because it is too large
to travel between cells. The search continued,
and five years later Fleming found his true
antibiotic drug: *penicillin*.

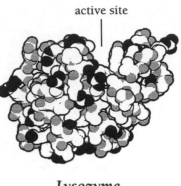

active site

Both lysozyme and penicillin destroy the
protective cell wall of bacteria. Bacteria build a
tough skin of protein and carbohydrate, which
braces their cell membrane against a high internal
osmotic pressure, estimated at 5 to 30 times the
pressure of the surrounding air. Lysozyme breaks the
carbohydrate chains in this structural network and penicillin blocks the enzyme
that builds them. Either way, the bacteria lose their structural integrity and burst
under their own pressure.

Lysozyme

Places rich in potential food for bacteria are often protected with lysozyme.
Egg whites are rich in lysozyme, to preserve the proteins and fats that will nourish
the developing chick. Tears and mucus contain lysozyme to resist infection of our
exposed surfaces. The blood is the worst place to have bacteria grow, as they
would be delivered to every corner of the body. Lysozyme in the blood lends some
protection, but far more powerful methods, employed by the immune system, are
also brought to bear on internal infections.

α-Macroglobulin

Digestive enzymes, such as the serine proteinases (page
56), are sturdy enzymes. They are simple to build and
difficult to destroy. Invading bacteria produce digestive
enzymes, often in unwelcome places, to allow them to
infiltrate into the body. The α-macroglobulins, circulating in the blood, are our protection against these enzymes.
They trap digestive enzymes before they do significant
damage, packaging them safely for disposal. Bird and
reptile egg whites also contain significant amounts of α-
macroglobulin, to protect the proteins stored there.

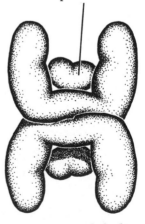

proteinase

α-Macroglobulin is a molecular mousetrap.
It is shaped like a letter H, formed from two identical

α-Macroglobulin

protein subunits. Between the two jaws is a flexible protein loop that acts as the bait. When it is clipped by a digestive enzyme, the entire protein snaps shut, trapping the proteinase inside. This change in shape also exposes a signal patch on the surface, which is recognized by cells that absorb and discard filled traps.

Metallothionein

Metallothionein is a small protein with a healthy appetite for heavy metal ions. The lighter metals, such as magnesium and calcium, are relatively easy to hold. In proteins such as amylase (page 61) and calmodulin (page 153), the oxygen atoms in glutamate and aspartate are strong enough to entrap them. Heavier metals, however, require stronger methods. Iron is often held inside a large planar heme molecule, as in hemoglobin (page 75), which surrounds the iron ions with four perfectly positioned nitrogen atoms. Sulfur atoms also form strong bonds with heavy metals, so cysteine is often used to entrap them. Metallothionein is unusual among proteins in that one-third of its component amino acids are cysteines (typical proteins may carry one or two cysteines in every 100 amino acids). The sulfur atoms in these 20 cysteines are hungry for metal ions—each metallothionein molecule may hold up to a dozen. The protein folds into a dumbbell shape. Each side has a cavity lined with cysteine sulfur atoms, forming two comfortable nests for heavy metals.

It is thought that metallothionein acts as a reservoir for important metals, such as zinc and copper, delivering them to sites of protein synthesis. New metal ions are in continual demand for building new metalloenzymes. In the liver, metallothionein also soaks up toxic heavy metals such as cadmium, lead, and

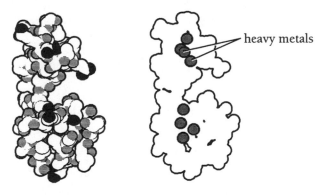

heavy metals

Metallothionein

mercury. It is made in large quantities when a person is poisoned with heavy metals. Heavy metals are dangerous because they bind tightly to cysteines in other proteins, often blocking their action. Metallothionein sequesters these metals where they can do no harm.

Immunity

Circulating in the blood is a powerful collection of molecules that recognize and destroy invading organisms. Collectively, they are known as the *immune system*. Because of them, we can go to bed when we get sick and wait for the body to destroy an infection. In some cases, we can even bolster the effect. Vaccination primes our immune system for possible future infection. Antibiotics help the immune system by weakening bacterial attackers. Today, however, the constant threat of AIDS has underscored the fragility of the immune system. Human immunodeficiency virus (HIV) is capable of disrupting the entire immune mechanism, leaving the body open to attack by countless infectious organisms.

White blood cells orchestrate the action of the immune system. Lymphocytes, one type of white blood cell, crawl through the body looking for foreign molecules. If they find something, they flood the area with *antibodies*, which promptly cover the foreign object. Other white blood cells seek out this bristling coat of antibodies and destroy whatever is inside. If the attacker is a bacterium, it may be destroyed by the *complement system*, which punches holes through its surface. Or it may be eaten whole by *macrophage* ("big eater") cells. We make a substantial investment in the immune system. More than two trillion white blood cells circulate through the body. If taken together, they would comprise an organ of about the same size as the brain. Protection is just as important to us as control and thought.

Antibodies (Immunoglobulins)

The heart of the immune system is the *antibody*, a protein that recognizes foreign molecules. Antibodies, also known as immunoglobulins, ignore our thousands of normal molecules. They seek out only the unique bacterial and viral molecules that accompany an infection. Antibodies provide the specificity for fighting disease. They tag unfamiliar objects so that we can distinguish them from our natural molecules. Antibodies circulate in the blood, searching each cell for parasites and infection. Of course, it would be disastrous for the immune system to build

Antibodies

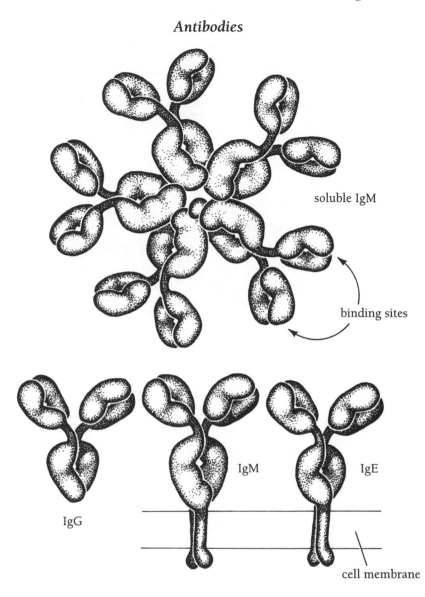

soluble IgM

binding sites

IgG

IgM

IgE

cell membrane

antibodies that recognize our own molecules. Antibodies would then coat normal cells, and the force of the immune system would be focused on normal tissues. The body would attack itself. (This actually happens in *autoimmune diseases*, such as rheumatoid arthritis.) This potential problem is solved in the thymus gland, where lymphocytes are born. Through an intricate recombination of genes, the thymus generates lymphocytes that recognize a wide range of possible molecules.

125

More than 95% of these cells, however, are destroyed, because they recognize molecules normally present in the body. The 5% that remain comprise the immune system, recognizing only foreign molecules—molecules that, if found, must be part of an infection.

Antibodies are Y-shaped proteins with a binding site for molecules at the tip of each arm. The binding sites are formed by six loops of the protein chain. In different antibodies, these loops differ greatly in shape and chemical composition, forming differently shaped pockets for binding to differently shaped molecules. Each lymphocyte builds only one type of antibody, which is tailored to bind to only one type of molecule. Some bind to small protein loops on the surface of a virus, others to unusual sugars on the surface of a bacterium. If a particular lymphocyte is called into use, its antibodies may be fine-tuned in the process of *affinity maturation*, improving them to bind even more tightly and specifically to their target. The lymphocyte mutates the amino acids in these six loops at thousands of times the normal rate, searching for combinations that improve the effectiveness of the antibody.

Antibodies are highly flexible: the arms may flap, bending at the connection between the stem of the Y and the arms, and each arm may twist and flex, bending at an elbow in the middle of the arm. This flexibility allows a single antibody to bind to two adjacent molecules on a foreign object, even if the distance between the molecules varies. Binding at two sites on a bacterium or virus, instead of just one, is a great advantage. Having two binding sites linked together increases the total binding strength, or *avidity*, not by a factor of two, but by over one thousand times. Two-armed antibodies stick tightly to their targets, planting tiny warning flags.

Variations on the basic Y-shaped antibody are built for different stages of protection: one for initially recognizing a molecule, another for destroying a foreign cell, and another for preventing future infection. Before ever encountering a foreign molecule, lymphocytes make *IgM* (immunoglobulin M). A short tail composed of carbon-rich amino acids anchors *IgM* to the outer surface of the lymphocyte. The arms of IgM extend outward from all sides of the lymphocytes, searching for foreign objects. When a lymphocyte finds something, the *primary immune response* begins. Since the binding of the antibody is not perfectly tuned, it is important to have many sites for binding, to increase its avidity. The lymphocyte floods the infected area with a soluble form of the IgM molecule, composed of five Y-shaped molecules connected together in a star. One or two weak binding sites may not be enough to attach an untuned antibody firmly to the target, but the 10 binding sites on IgM will.

The immune system has now realized that something is amiss. In the *secondary immune response*, which peaks about 10 days after the initial recognition of disease, the full force of the immune system is brought to bear against the infection. Lymphocytes build massive quantities of fine-tuned *IgG*, the archetypal antibody, which coat every unnatural surface in sight. About three-quarters of the antibodies circulating in the blood are IgG. When IgG binds to a foreign cell, both the complement system (page 130) and phagocytes ("eating cells") battle the infection.

Immunity to a disease is achieved after an infection is conquered, when we know how to destroy a particular pathogenic organism. We save the lymphocytes that made the proper antibodies, leaving the immune system in a state of readiness for a second attack by the same organism. *Vaccination* is a medical way of fooling the immune system into doing the same thing, achieving immunity before an infection. The polio vaccine given to a child is a disabled version of the polio virus, strong enough to activate the proper lymphocytes, but not strong enough to cause polio. Soon after our vaccination, we destroyed these weak viruses. Today, the lymphocytes from this minor skirmish still circulate through our blood, ready to battle a future attack by the real virus.

The major cause of allergies is the antibody *IgE*. Like the IgM on the surface of lymphocytes, IgE has a short membrane anchor on its stem, affixing it to the surface of white blood cells. When IgE locates a foreign molecule, like the molecules on a pollen grain, it triggers the cell to release histamine (page 142) instead of more antibodies. Histamine initiates the miserable symptoms of an allergic response. The runny nose, sneezing, and watery eyes are designed to flush out the offending molecules. An *antihistamine*, as the name implies, blocks this signal.

MHC (Major Histocompatibility Complex)

Viruses are insidious attackers. When a cell becomes infected by a virus, the infection may remain completely hidden inside. The virus hijacks the cellular machinery, forcing it to make thousands of new viruses. The damage, however, is done while the virus is safely sealed inside the cell membrane, invisible to anti-

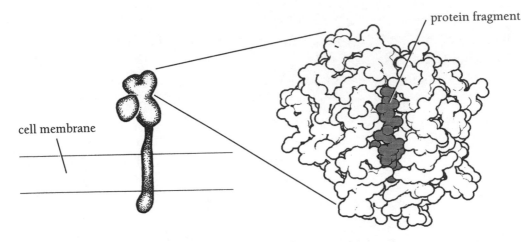

Major Histocompatibility Complex

bodies. Only too late, when the mature viruses burst from the cell, is the infection apparent. How can our immune system attack something it cannot see or find? We have an elegant solution to this problem: infected cells signal the immune system using the MHC (major histocompatibility complex). An infected cell, as it attempts to defend itself, breaks some of the rapidly proliferating viral proteins into pieces. The cell then presents these pieces on its surface, displayed prominently inside a groove at the top of the MHC. The MHC extends from the cell surface like a flagpole, flying its viral protein fragment. Lymphocytes easily see these fragments, using *T-cell receptors* on their surfaces, and signal to the immune system to destroy the cell, growing viruses and all.

The MHC, as its name implies, is largely responsible for the acceptance or rejection of skin grafts and organ transplants. The donated tissues carry the donor's MHC, which is usually different from the patient's own MHC. The patient's immune system gets confused. The new MHCs are the wrong shape, so the patient's lymphocytes interpret them as belonging to an infected cell and the graft or transplant is destroyed. Thus, it is imperative to match the MHCs of the donor and the patient. Our reaction to poison ivy is also initiated by the MHC. The process of sensing the offending plant protein, activating white blood cells, and finally cleaning up the poison takes several hours—the immune system is not fast, but it is painstakingly thorough—so the rash and itching develop long after the walk in the woods.

CD Proteins

CD proteins (CD is shortened from cluster of differentiation, a term derived from the genetics of this group of proteins) act as "molecular Velcro," strengthening the tenuous connections between cells of the immune system. For example, an infected cell may display only a few MHC molecules containing viral fragments. Lymphocytes must be able to bind tightly to this unfortunate cell in order to have an effect, but the interactions between T-cell receptors and these few filled MHC proteins will form only a weak link. This is where CD enters the picture. After a few key T-cell receptor–MHC connections are made, bristling CD molecules on the lymphocyte adhere to empty MHC proteins in the vicinity, stitching the cells firmly together.

CD4 and *CD8* extend from the surface of lymphocytes, searching for different forms of MHC. Both are quite flexible, allowing them to search for their targets. The elongated body of CD4 is flexible at a central hinge, with the binding site for MHC at the tip. Notice that the body is formed of two similar comma-shaped segments arranged in tandem. It is thought that CD4 evolved by gene duplication: the DNA instructions for one half were copied and spliced after the original, forming new instructions for a protein twice as large. CD8 is attached to the cell surface by two long, flexible ropes, making it even more flexible than CD4. Bulky carbohydrate groups along the rope prevent the formation of a normal, compact protein structure, keeping the chain extended and supple. The small globular head swinging on this rope, similar to the tip of an antibody, attaches to MHC.

HIV prefers to infect T lymphocytes (lymphocytes from the thymus), and uses CD4 to find them. The surface of HIV is studded with *GP120* (a glycoprotein with a molecular weight of about 120,000 times that of hydrogen), which binds

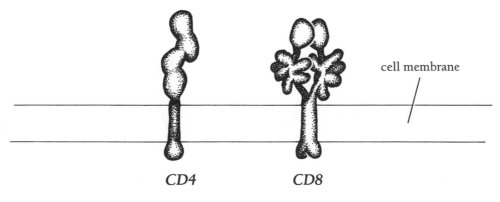

cell membrane

CD4 *CD8*

tightly to CD4. This is the first step of infection by the virus, affixing the virus to the lymphocyte surface. One approach being tested to fight infection by HIV is to administer a water-soluble form of CD4: CD4 that has been cut away from its membrane anchor. Soluble CD4 binds to GP120 on free viruses, blocking their grappling hooks and making them unable to bind to a lymphocyte. Of course, as a side effect, the soluble CD4 will also block the normal functions of CD4.

Complement

The complement system is our primary defense against bacterial infection. After antibodies have found a bacterial cell, the complement proteins recognize the antibodies coating the surface and initiate a bactericidal *cascade*. Cascades are molecular chain letters: one molecule starts the chain by sending a message to several individuals of the second type. Each of these then activates several individuals of the third type. And these on to a fourth type. Each step activates an increasingly large circle of molecules, just as a chain letter builds an increasingly large circle of addressees at each mailing. A single molecule can ultimately activate tens, hundreds, or thousands of molecules at the final step. In the complement cascade, the weak signal from a single bound antibody is amplified through several steps to create about a dozen *membrane attack complexes*, which pierce holes in the invading bacterium.

The trigger for the complement cascade is the protein *C1*. It has the distinctive look of an immune system protein: several flexible arms with binding sites at the tip of each. Molecules of the immune system cannot afford to be overly rigid—they must be able to adapt to the arbitrary shape of an invading organism. C1 searches for antibodies bound to a foreign particle. When several of its arms bind simultaneously, as when they find a star-shaped IgM stuck to a bacterial surface, C1 changes shape, beginning a series of cleavage reactions. Two proteins at the top of C1 are *serine proteinases* (with an active site similar to the digestive serine proteinases; page 56) that are normally in an inactive form. After binding to antibodies, however, they become active proteinases, triggering the first step in the

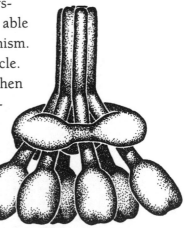

Complement C1

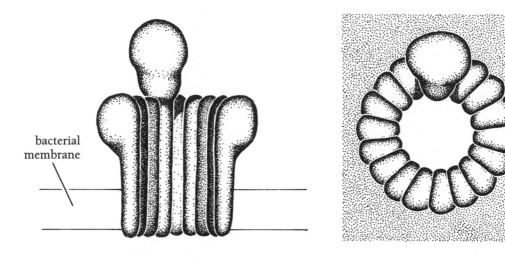

Membrane Attack Complex

cascade. They activate the proteins C2 and C4 by making specific cuts in them. Together, active C2 and C4 then snip a few pieces off C3, activating it. Activated C3 then performs the final step of the cascade, triggering first C6, C7, and C8 to penetrate the bacterial membrane, then triggering many C9 molecules to insert on either side, forming a tubular membrane attack complex.

The hole formed by the membrane attack complex kills a bacterial cell not so much by allowing important molecules to leak out as by allowing water to flow in. Since the concentration of ions, proteins, and other molecules is much higher inside cells than out, water rushes inside when given a chance. Normally, cells carefully control the influx and efflux of water, but the membrane attack complex circumvents this control, allowing water to flood in. The bacterium bursts under the pressure, and white blood cells clean up the debris.

Healing

When we suffer a wound, it is imperative to block the leak as quickly as possible. Blood is a valuable resource with many roles: defense against disease, transport of nutrients, delivery of messages. But it is a liquid resource easily lost through a simple wound. Blood is pumped rapidly and at high pressure: our 5 liters of blood pass through the heart once every minute. An emergency repair system is critically important in case of an accident. Circulating along with the other components of the blood, standing ready, is the machinery needed to erect a temporary dam. In a matter of seconds, a blood clot may be built to block any leak, ensuring that the rapid flow of blood remains inside where it is needed.

Blood clotting is a delicate business. Clots must be carefully localized. If any small cut on the finger caused all of the blood in the hand to solidify, the entire circulatory system would grind to a halt. Clots must form only at the site of a cut, blocking only the severed vessels. We achieve this precise control through the combined action of dozens of proteins—termed blood clotting factors—and the *platelets*, tiny fragments of cells formed in the bone marrow. Together, circulating platelets and blood clotting factors build a plug at a site of damage, then build a stronger dam, then dissolve away after the wound has healed.

Platelets are assigned the task of initially recognizing a wound. Normally, they only make contact with the inside of vessels as they circulate in the blood. At a wound, however, platelets flow to the outer surfaces and find the many structural molecules surrounding the vessels, such as collagen (page 95). These molecules signal that there is a problem. Platelets attach to the surface of the cut and immediately signal to other platelets to do similarly. Through a surprising economy, the signal is a concentrated burst of ADP (page 67), using the molecule of energy transfer as a simple messenger. Receiving this message, platelets stack like bricks onto the initial layer, forming a fragile cellular plug. This is only a stopgap measure; concurrently, the blood clotting machinery is activated. In a cascade of action, strong fibers of protein are built around and inside the platelet plug, which dries to form a tough red scab that protects the wound as it heals.

von Willebrand Factor

Flexible strands of von Willebrand factor act as the bridge between platelets and the surface of a wound. (This factor is named after the doctor who first described a case of hemophilia caused by the lack of this protein.) Von Willebrand factor forms tangled aggregates in the blood, with long, dumbbell-shaped molecules bound end to end in an extended string. The larger portions at the ends of the dumbbell are thought to bind to molecules such as collagen; the long rod-shaped center portion is thought to bind to the surface of platelets. Imagine platelets stickily squashing a mat of von Willebrand factor against the raw surface of a cut, like stepping onto gum on a warm sidewalk.

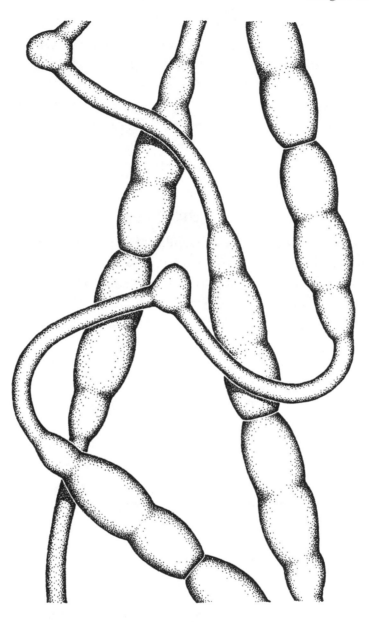

von Willebrand Factor

Upon binding, platelets send a flurry of signals. A concentrated burst of ADP (page 67) signals to other platelets, causing them to cluster into the wound, helping to block the flow of blood. The small, lipidlike molecule *thromboxane A$_2$*

signals to the bloodclotting machinery, beginning the cascade that leads to a tough fibrin clot (see below). Thromboxane A$_2$ is formed by the enzyme *cyclooxygenase*. Cyclooxygenases are a major target of *aspirin* (acetylsalicylic acid). Small doses of aspirin inhibit the cyclooxygenase of platelets, but larger doses are needed to stop the similar cyclooxygenase that makes the molecule that signals pain (page 143). Small, repeated doses of aspirin are enough to block the cyclooxygenase in platelets, slowing the formation of blood clots. This is beneficial in people at risk for stroke or heart attacks. Aspirin inhibits the formation of clots in the blood vessels nourishing the brain and the heart.

Blood Clotting Factors

A cascade of blood clotting factors, similar to the complement cascade (page 130), amplifies the signal from a wound into a tough fibrous clot, strong enough to last the weeks needed to heal a wound. The cascade starts when *tissue factor*, also known as thromboplastin or factor III, is exposed to blood. (Many of the clotting factors were discovered independently by several researchers, leading to a confusing array of different names.) Tissue factor is the only factor not normally present in the blood. It waits outside blood vessels, on the surface of cells, in case of damage. When tissue factor comes in contact with blood, it activates *factor VII*, triggering the cascade. Factor VII then activates *factor* X, which in turn activates *thrombin*. Thrombin plays the central role, by activating the *fibrin* (see below) that actually forms the structural cables of the clot. At each of these steps, tissue factor–factor VII–factor X–thrombin–fibrin, a single protein activates many copies of the next protein. The signal of a few tissue factor molecules is amplified through the cascade to form thousands of active fibrin molecules.

Blood Clotting Factors

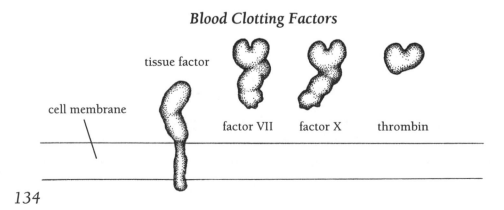

tissue factor

cell membrane

factor VII   factor X   thrombin

This cascade must be carefully contained, so that the wave of clotting does not spread to the neighboring healthy tissues. Control is accomplished in several ways. Many of the factors have very short lifetimes once they are activated. The inactive form of thrombin lasts for hours in the blood, but when activated, it last only seconds. Thrombin can cover only a short distance in its short active life. Several other factors, such as tissue factor, are active only when bound to a cell surface, so they cannot spread through the blood at all. Thrombin also locks itself in place, by actually sticking to the fibrin strands in the clot, remaining exactly where it is needed. And on top of these physical limitations, an array of activators and inhibitors modulate each step.

Vitamin K, obtained primarily from green vegetables and from bacteria in the intestine, is essential for the synthesis of several of these factors. They are made like normal proteins, but then must be modified for proper action. Extra negatively charged groups are added onto their surfaces to enhance their interaction with calcium ions. The enzyme performing this chemical modification requires vitamin K to assist in the reaction. The rat poison *warfarin* blocks the action of vitamin K and thus acts as a powerful anticoagulant. Anticoagulants are also made, as one might expect, by organisms that feed on blood. The black medicinal leeches used throughout history secrete *hirudin*, a small protein that binds tightly to thrombin. By blocking the action of thrombin, the leech slows the formation of a clot, allowing more time to obtain a meal.

Hemophiliacs tend to bleed more profusely and for longer times than most people because they lack one of these blood clotting factors. The classic historical hemophilia of European nobility, and many people today, is caused by the absence of factor VIII, an accessory activating protein. Other, rarer forms of hemophilia are caused by the deficiency of one or another clotting factor. Hemophilia can be treated by injection of the factor purified from blood. Before the development of accurate tests for screening blood, however, this treatment carried a heavy price: possible exposure to HIV.

Fibrin

The visible structure of a blood clot—the thick red glue that dries into a hard scab—is composed of blood cells caught in a web of fibrin fibers. Fibrin is built as inactive *fibrinogen*, which quietly circulates throughout the blood. At the site of a wound, it is activated by thrombin, which clips off small pieces at the center.

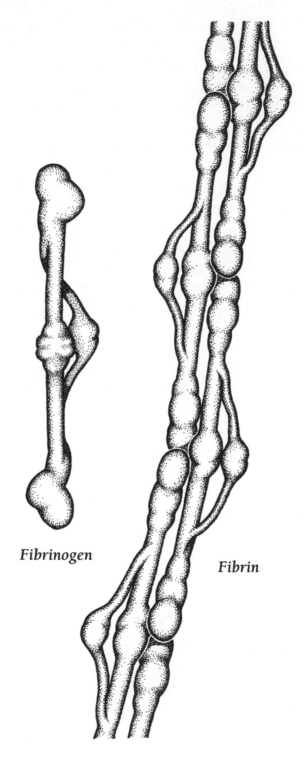

Fibrinogen

Fibrin

This opens up a sticky site, and the slender fibrin molecules bind side by side, assembling a tough, double-stranded cable. These cables branch and associate with neighboring strands, forming a tangled web. When the strands are in place, crosslinks lock them into a rigid structure that traps red blood cells like fish in a net.

After blocking the flow of blood from a wound, fibrin clots act as a scaffolding over which the healing tissue can construct new tissue and blood vessels. As the wound is closed, the scaffolding must be removed while new architecture is constructed. *Plasmin*, a small digestive protein, disassembles the scaffold. It cleaves fibrin filaments primarily in the rod-shaped segments between the central knob and the outer domains, neatly breaking the fibrin fibers into small discardable fragments.

6
Molecules and the Mind

"... all music is but three parts vied,
and multiplied"   *George Herbert*

The greatest wonder of the molecular world is that it gives rise to the manifold subtleties of sense, thought, and emotion. One might easily imagine tiny molecular machines digesting food or molecular ropes and ratchets powering our arms and legs. These processes are inherently physical, and similar in many respects to processes in our familiar world. But the hazy memory of a childhood friend is far more difficult to imagine at the molecular level. It is becoming clear, however, that each aspect of our inner selves may be traced to a physical, molecular source. Hunger, imagination, pain, sight, fear, dreams, anger, smell, inspiration, lust, touch, taste—all have their roots in the multiple actions and interactions of molecules. Just as a touch of the pendulum endows a collection of gears and weights with the ability to measure time, so too by setting the last molecule in place does a new life begin to think and feel.

The body sends and receives many different types of messages. Each feels different, yielding the varied manifestations of thought and emotion. The body sends messages to itself, without any need for conscious intervention. Cells talk to their neighbors about overcrowding, the pancreas speaks to the liver about sugar levels, the genitals tell the body when it is time to mature. *Pain*, in contrast, is a message that demands instant, personal response. Pain is a message, sharp and insistent, sent by the body to the brain, indicating that something is amiss. The *senses* are messages that the eyes, ears, tongue, nose, and skin send to the brain. The senses translate external messages—light, pressure, heat, fragrance—into the language of the nervous system. *Emotions* are messages sent by the brain to the body, spreading like a wave to reach every cell. These deeply internal messages control shifts of mood, or heighten our responses in times of emergency. *Thoughts* are messages that the brain sends to itself, interpreting the messages from the senses and weighing responses from minute to minute. If action is necessary, the brain dispatches a command to the muscles. In the very best of times, however, these messages may never leave the brain at all, giving rise to our inner world of ideas and imagination.

Messages are sent throughout the body in two distinct manners. The simplest, but most wasteful, messages are written in molecular letters, known as *hormones*, which are delivered through the bloodstream. Hormones are built in one part of the body, dropped into the blood, and picked up and read at a distant part. The shape and size of each hormone contains the text of the message, for the surface of a hormone is read like braille by the receiving cell. This type of message can be quite costly, as each new message requires the construction of an entirely new molecule, with a new size and shape. We reserve these messages for the simplest and most basic messages, messages involved in body maintenance and messages rooted deep in our emotions, like "I'm thirsty!" or "Help!"

For more complex messages, we employ a more versatile system of communication. These messages are transmitted down nerve fibers, like electronic messages traveling down a wire, more rapidly and with less waste than the physical hormone messages in the blood. Communication through the nervous system is more akin to a telephone conversation than a letter in the mail. It is also infinitely more flexible. The true power of the nervous system is the ability to reconnect nerves in new configurations, like a telephone switchboard, building new networks to respond to each new need. Nerves deliver the messages of motion, orchestrating the contraction and release of muscles; the information from the senses is collected, sorted, and processed by nerve signals; and the interplay of nerve signals inside the brain gives rise to the loftiest of messages: conscious thought.

Hormones

Hormones are our most basic molecular messages. Hormonal messages may not be consciously felt at all. The minute-by-minute maintenance of blood pressure and heart rate, the timing of digestive motions, and the control of sugar and fat delivery through the blood are all regulated through a flurry of hormonal messages, with no conscious help from ourselves. Other hormonal messages are felt deeply, not as conscious thoughts, but as emotions. The flush of fright when we are startled is the spread of epinephrine, priming us for action. Hormones are constructed in glands—the pituitary, the hypothalamus, the adrenal glands, the genitals—and dropped into the bloodstream for delivery. Their effects are global, affecting tissues throughout the body.

Nitric Oxide

Nitric oxide, a gas composed of a single nitrogen atom paired with a single oxygen atom, is our smallest hormone. Nitric oxide is a reactive compound, useful in the body but destructive in the environment. In automobile engines, it is made when nitrogen gas and oxygen gas are forced together under high pressure and heat, forming a powerful pollutant that chemically destroys ozone in the upper atmosphere. We produce nitric oxide less violently, from oxygen gas and arginine by the enzyme *nitric oxide synthetase.*

Nitric oxide in large amounts is toxic. (The similar compound nitr*ous* oxide, with two oxygen atoms and one nitrogen atom, is less toxic—it is the anesthetic known as "laughing gas.") Nitric oxide combines with iron atoms in proteins, much as cyanide and carbon monoxide do, blocking their action. The immune system takes advantage of this toxicity, producing a localized burst of nitric oxide to kill infected cells or tumors. When acting as a messenger, however, nitric oxide is used in trace amounts. It binds to iron atoms in the receptor proteins that receive the message. Nitric oxide controls the dilation of blood vessels, coaxing vessels to expand. Constriction of blood vessels in the heart can cause severe chest pains, which are treated in minutes with a tablet of *nitroglycerin* dissolved under the tongue. Traveling in the blood, nitroglycerin releases nitric oxide, signaling blood vessels to dilate, reducing blood pressure and alleviating the problem.

Histamine

When a cell becomes damaged, perhaps due to a cut or abrasion, it sends a cry for help. Damaged cells release a wave of histamine, signaling trouble. In response, blood vessels in the area dilate, allowing more blood to enter and cleanse the area. The tissue reddens with the extra blood flow, becoming *inflamed*. This signal occasionally backfires. Cells faced with a minor irritant, such as pollen, think

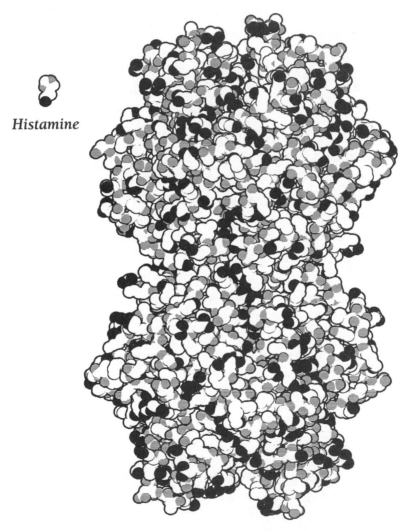

Histamine

Histidine Decarboxylase

they are in life-threatening danger and release histamine, causing the symptoms of allergy and asthma. Hornets and nettles inject a mixture that includes histamine, which rapidly inflames the area. This painful reaction is not due to any particularly nasty properties of the histamine molecule. The extreme pain is due to the way it is seen as a signal—histamine in this excessive quantity is heard as a piercing scream.

Histamine is made from the amino acid histidine by the enzyme *histidine decarboxylase*. It removes the acidic group from histidine, forming carbon dioxide gas and histamine. Histamine decarboxylase requires a helper molecule to perform this extraction. Our enzyme uses vitamin B_6 to provide the necessary chemical leverage. *Antihistamines* block the action of histamine, reducing inflammation and allergies. Most antihistamines act not by stopping the formation of histamine—this might be dangerous, because histamine has many roles in normal tissue growth and repair—but instead by selectively blocking the site that receives the message. Diphenhydramine (the active ingredient of Benadryl) and chlorpheniramine (the active ingredient of Chlor-Trimeton) are small molecules that superficially resemble histamine. They bind tightly to histamine receptor proteins on the target cells and block the action of histamine. The unpleasant affects of motion sickness are also caused by histamine, and are easily blocked by the antihistamine dimenhydrinate (Dramamine).

Prostaglandins

Injured cells ask for help from their neighbors by sending histamine, which then respond by allowing protective blood cells to enter. At the same time, injured cells also dispatch a message to the brain, demanding that something more permanent be done about the problem. We respond by dropping the hot pan or jerking away from the sharp edge of the knife. These messages are felt as pain and are sent by prostaglandins. Cells in distress release a wave of prostaglandins, which sensitize local pain receptors, sharpening their signals to the brain.

Prostaglandins are members of a large class of similar hormones collectively termed the *eicosanoids*. Other examples include thromboxanes (see page 134), which are important in the clustering of platelets in blood clotting, and leukotrienes, which carry some of the many signals between immune system cells. All are made from the 20-carbon tails of phospholipids (page 82). First, a single tail is cleaved from the lipid by phospholipase A_2, similarly to the phospholipases used in digestion (page 63). Then, two oxygen molecules are added by *cyclooxygenase*. Finally, one of the oxygen atoms is removed by *peroxidase* to form the prostaglandin. These latter two enzymes associate with one another to form a complex, often termed *prostaglandin synthase*. The complex attaches to a lipid membrane, with the deep cyclooxygenase active site

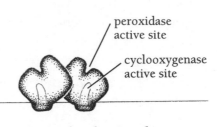

peroxidase active site

cyclooxygenase active site

Prostaglandin Synthase

dipped into its greasy interior. The enzyme extracts the carbon-rich tails, which are insoluble in water, directly from the membrane. After addition of oxygen, the tails are quite soluble, and easily find their way to the peroxidase active site on the opposite side of the complex.

Analgesic drugs are available to block nearly every step of pain transmission. Steroid analgesics, such as cortisone, block the phospholipase that originally releases the carbon-rich tail. Nonsteroidal antiinflammatory drugs (NSAIDs), such as aspirin and ibuprofen, bind in the deep cyclooxygenase active site, blocking its reaction. Opiate drugs, such as morphine, act not at the source but at the receiving end, blocking the nerves that sense and interpret pain signals.

Hypothalamus/Pituitary Hormones

At the center of the brain, there is a portal of communication between the realm of mind and the realm of body. There, the hypothalamus translates nerve signals into the hormonal language that the body understands. When the brain signals, the hypothalamus triggers the pituitary gland, which promptly releases a variety of hormones into the blood for delivery to distant parts of the body. These hormones are typically composed of small protein chains, making them easy to build and easy to discard.

Oxytocin and *vasopressin* are very small proteins, with nine amino acids each. Oxytocin delivers a specialized message for women with children. It begins the muscle contractions of birth, and it later triggers muscles around the milk ducts to contract, releasing milk. Vasopressin, also known as antidiuretic hormone, differs from oxytocin in only two of its nine amino acids: a small isoleucine is changed to a bulky phenylalanine and a small leucine is changed to a charged arginine. This small chemical difference is read as an entirely different message. Vasopressin tells the kidney to reabsorb water, serving as our natural antidiuretic.

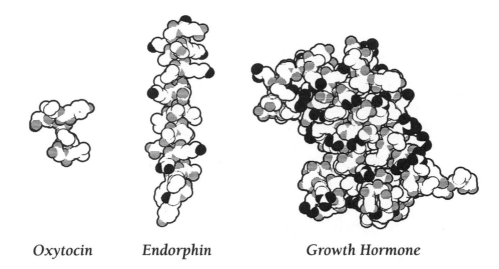

Oxytocin   *Endorphin*   *Growth Hormone*

The *endorphins* (*endo*genous m*o*rphinelike hormones) and *enkephalins* modulate the perception of pain. They are small proteins of 5 to 30 amino acids. Under their action, pain is still felt but is not perceived as unpleasant. *Morphine* mimics these hormones, fooling the brain into deadening the pain. As one might expect from a drug that acts directly on the brain's natural painkilling mechanism, morphine is the strongest painkiller known. Morphine also blocks the nerves responsible for the reflex of coughing, but it is far too narcotic for this use. However, the similar molecule *codeine* is often used. Codeine has a much weaker painkilling effect and is thus less narcotic, but it is similar in effectiveness as a cough suppressant.

Growth hormone is considerably larger, at 191 amino acids, but is still small for a protein. As its name implies, it triggers increased growth. During childhood, a deficiency of growth hormone can lead to dwarfism and excessive amounts can lead to gigantism. Treatment with growth hormone during childhood can reverse the effects of low natural levels, restoring normal growth. Previously the hormone was isolated from pituitary glands of cadavers, a very scarce source at best, carrying also the possibility of viral contamination. Now the hormone is produced safely and in large quantities by engineered strains of bacteria. Growth hormone, like the steroid hormone testosterone (page 149), also increases general levels of protein synthesis. Therefore, athletes have experimented with both hormones in an attempt to increase muscle bulk.

Growth Factors and Cytokines

Cells are in constant communication with their neighbors, to police the proper growth of their local neighborhoods. Some tissues must grow continually: fibroblast cells must replenish the skin, and bone marrow cells must continually build new blood cells. Other tissues grow only rarely, to replace old cells or repair damaged areas. An array of protein hormones, termed *growth factors*, carry messages between our cells, giving permission to grow or warning that an area is too crowded. Cells receive these messages with receptor proteins (page 151) that are arrayed on their surfaces. Receptor proteins wait for growth factor messages; after receiving them, they signal inside the cell, mobilizing the machinery for growth and reproduction. The genes for growth factor receptors are often called *oncogenes*, because of their intimate linkage with cancer. If an oncogene is mutated, perhaps by ultraviolet light or carcinogenic chemicals, the receptor protein it builds will not be active. The cells with the defective gene

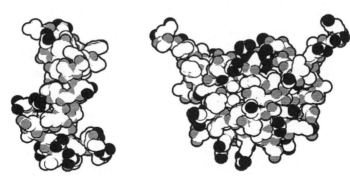

Epidermal Growth Factor   *Interleukin-8*

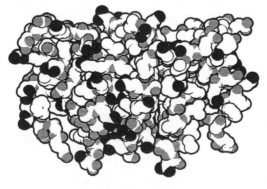

Interferon-γ

146

lose their normal controls on growth; they are free to divide continuously, forming a cancerous growth.

Cells in the immune system must also stay in constant communication with one another, in order to mobilize defenses when they detect a foreign organism. White blood cells communicate using small protein messages termed *cytokines*. They are released into the blood, quickly spreading the message to the surrounding cells. A series of cytokines termed *interleukins* (abbreviated IL) announce each step of defense against infection. IL-1 signals that an attacker has been found, activating T lymphocytes in response. IL-2 informs T lymphocytes to remain activated over the time it takes to combat the infection. IL-4 tells B lymphocytes to switch into their full antibody-producing mode. IL-8 is a beacon at the site of infection: white blood cells near an infection release IL-8, which acts as an attractant, coaxing other cells to the area.

Interferons are cytokines involved in the destruction of viruses. They tell the immune system to focus on cell-mediated approaches, pouring resources into the cell-killing ability of cytotoxic T lymphocytes or the cell-eating ability of macrophages instead of antibody approaches. Interferons also tell infected cells to build virus-fighting molecules. Interferons initiate the same responses against cancer cells, mobilizing immune system cells to fight the unnatural growth. Although severe side effects limit the doses that may be administered, interferons are being used to fight some cancers, such as Kaposi's sarcoma, and some viral diseases, such as warts.

Insulin and Glucagon

Blood sugar level is regulated by the opposing actions of two hormones: insulin and glucagon. Insulin warns that the sugar level is too high, glucagon that it is too low. The pancreas continually tests the blood, releasing one or the other based on the results. The signal is received primarily by the liver, which either stops releasing glucose into the blood or increases the amount. This minute-by-minute monitoring of levels of glucose is particularly important for feeding the nervous system, which relies almost completely on blood-borne glucose for energy.

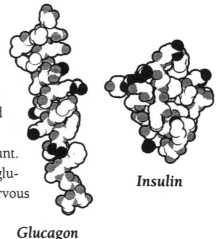

Insulin

Glucagon

Insulin deficiency, due either to genetic deficiency or to changes brought on by age, leads to *diabetes mellitus*. The lack of insulin allows sugar levels to get too high after meals. The condition is treated by injecting insulin at appropriate times, manually performing the job of the pancreas. Insulin must be injected; because it is a protein, it would be rapidly digested if taken orally. Insulin from cows and pigs is commonly available. It is similar enough to human insulin to fool our cells. Actual human insulin, made by engineered strains of bacteria, is also becoming more widely available.

Thyroid Hormones

The thyroid gland, located in the neck, produces the hormones *thyroxine* and *triiodothyronine* from the amino acid tyrosine. These hormones speak to nearly every one of our cells, regulating our normal level of metabolism. They orchestrate the amount of oxygen and sugar needed from day to day. They balance the breakdown of resources with the need for new growth, and maintain our optimal body temperature. The thyroid gland has a unique requirement for *iodine* to construct these thyroid hormones. Our natural source of iodine is from saltwater fish and algae. Organisms living in the ocean concentrate the trace amounts of iodine dissolved in seawater. Traditionally, people living far from the oceans did not have access to this source, and risked developing *goiters*. Iodine deficiency causes the

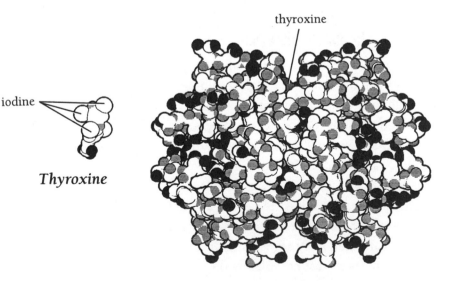

Thyroxine

Transthyretin

thyroid gland to swell until it is visible as a bump on the neck. Today, goiters are relatively rare, as a small amount of an iodine salt is commonly added to "iodized" table salt.

Many small hormones, including the thyroid hormones and the steroid hormones, and some vitamins, such as A and D, are not soluble in water. They do not dissolve in blood and must be shuttled around by specific carrier proteins, individually picked up in one part of the body and deposited elsewhere. The protein *transthyretin* transports the thyroid hormones, and similar proteins transport other carbon-rich hormones. The hormones are slowly released in the tissues, where they cross the cell membrane (being rich in carbon groups, they slip easily through cell membranes) and deliver their message. These insoluble hormones, inside their protein taxicabs, tend to circulate in the blood for much longer than water-soluble hormones like glucagon and insulin. Therefore, water-soluble hormones are often used for messages that are immediate in action, like the minute-by-minute adjustment of blood sugar levels. The water-insoluble hormones are used for longer-range messages, like "We have plenty of energy" or "We're pregnant."

Steroid Hormones

Steroids are small, carbon-rich hormones made from cholesterol (page 82). The *adrenal steroids* are made by the adrenal glands on top of the kidneys. The most familiar is *cortisol* (hydrocortisone), a hormone that stimulates the storage of glucose in the liver and enhances the breakdown of protein in muscle. Cortisol tells us to mobilize the resources locked up in protein and store them in a more readily usable form: liver glycogen. Cortisol also blocks the normal processes of inflammation and wound healing. Cortisol and several chemical derivatives are used topically to reduce inflammation of rashes and insect bites, and more invasively by injecting them into arthritic joints. These treatments do not cure the condition. They merely serve to reduce the natural, but often painful, responses of the body to damage.

Sex hormones are made in the ovaries and placenta of women and the testes of men. The major sex hormones in women are *estrogens* and *progestins*, and in men the major sex hormone is *testosterone*. The sex hormones are intimately

Steroid Hormones

estradiol

testosterone

involved in the physical differences between men and women. During the development of an embryo, they guide the formation of male or female genitals. At puberty, they give women their breasts and begin the menstrual cycle, and give men their beards and their larger bones and musculature and begin the production of sperm. In an adult woman, estrogens time the monthly cycle of fertility and progestins chaperone the changes of pregnancy. The libido of adult men is thought to be linked to levels of testosterone. In a remarkable irony, these most visible differences in physical appearance—leading to much of the world's art, literature, and day-to-day thought—are controlled by the presence or absence of a single methyl group. The two methyl bumps on testosterone mean "man," the single methyl bump on estradiol (an estrogen) means "woman."

Sex hormones are used in a number of artificial ways to modify the sexual characteristics of adults. Birth control pills are a mixture of estrogen and progestin. The pills disrupt the normal cycles of these hormones and thus prevent the signal to ovulate from being received, so that an egg is not released. Testosterone is widely used in sports to increase muscle mass and strength—in competition, officials often test for these "steroids." In the large doses taken, testosterone can have a number of unwelcome side effects, including induction of male characteristics in female athletes, impotence in men due to the disruption of the body's normal sexual signals, and profound psychological disturbances such as depression or mania.

Renin

Peptides—small proteins composed of less than a few dozen amino acids, such as oxytocin, vasopressin, and the endorphins—are used extensively as hormones. One might envision two ways of making a peptide of, for example, 10 amino acids. We could make nine different enzymes, each designed to add one additional amino acid to the chain. Or we could make a peptide with the normal protein synthesis machinery, according to a code in DNA. We actually use both approaches. For very small peptides, such as the three-amino-acid glutathione (page 118), a separate enzyme adds each part. Longer peptides, however, are made just like other proteins. Unfortunately, for such short lengths, the starting and stopping of ribosomes (page 49) is not overly successful. So, a longer protein is made, perhaps 100 amino acids, and then cut to size. The hormone *angiotensin*, which is important in regulating blood pressure and fluid balance, is an example.

It is built as *angiotensinogen*, a large protein of about 500 amino acids. When the active hormone is needed, the enzyme *renin* clips off a 10-amino-acid piece, which is then trimmed to 8- and 7-amino-acid lengths by other proteinases, forming the active hormone. Notice the resemblance of renin to the aspartyl proteinases made for food digestion (page 59). They differ, however, in that digestive enzymes break nearly every type of protein chain, whereas renin breaks only the single correct bond in angiotensinogen.

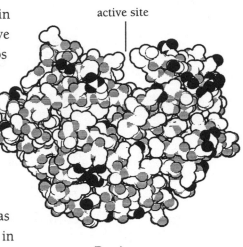

active site

Renin

Drugs that block renin are effective *antihypertensives*—drugs that reduce high blood pressure. They inhibit the maturation of angiotensin, removing the signal to increase blood pressure. Other antihypertensives act differently, fixing the symptom rather than the source. Some are diuretics that cause the body to lose water, thus reducing the volume of blood and reducing blood pressure. Other drugs, such as *minoxidil* and *cloniprin,* cause the thousands of tiny vessels throughout the body to relax and dilate, providing more room for the overpressured blood.

Hormone Receptors

Hormones circulating in the blood must somehow deliver their message across the seamless barrier of the cell membrane into the interior of a cell. Small, carbon-rich hormones, such as thyroxin and the sex hormones, dissolve through cell membranes, delivering their messages directly. But peptide and protein hormones, such as oxytocin and insulin, do not themselves enter into cells. Instead, their messages are relayed across the cell membrane by specific *receptor proteins*. Three entirely different methods are commonly employed to relay a message across the membrane.

The first type of hormone receptors create a pore through the cell membrane, similar to the pores created by porin (page 85) or connexons (page 87). In the absence of hormone, these pores are tightly shut, but when hormones bind to their outer surface, the pores open and allow ions to enter the cell. These recep-

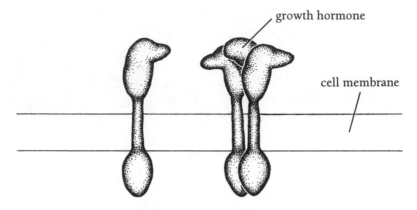

growth hormone

cell membrane

Growth Hormone Receptor

tors are commonly termed *gated channels*. They are typically very fast, snapping open and allowing tens of thousands of ions to rush inside. The ions then flow throughout the cell, delivering the message. The best-characterized gated channels are receptors for neurotransmitters, such as the acetylcholine receptor (page 163).

A second class of receptor uses an entirely different concept: the hormone assembles an active enzyme inside the cell. These hormone receptors are dumbbell-shaped, with one end inside the cell and one end outside. The clever trick is that two of these receptors bind to each hormone. When two receptors bind to a hormone outside the cell, their inner portions are brought together and form an active enzyme. These enzymes typically attach phosphates to proteins inside the cell, causing a conspicuous change in their shape and charge. Our receptors for insulin, growth hormone, cytokines, and a host of growth factors act in this manner.

A third class, including our receptors for epinephrine, glucagon, and vasopressin, initiates a cascade of molecules that center around the *G proteins*. Like the cascades of complement (page 130) and blood clotting (page 134), this cascade amplifies the small signal from a handful of hormone molecules into a significant cellular event. The receptor for the hormone is a large protein embedded in the cell membrane. On its outer surface is a pocket designed to accept a specific hormone. When a hormone binds, a change in shape propagates to the inner side of the receptor. There, it is recognized by the G protein, which separates into two pieces in response to the signal. The large piece then activates a cascade of proteins inside the cell. In some cases, the cascade results in the formation of

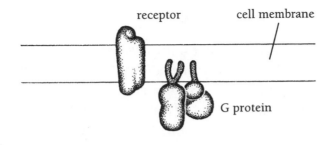

G-Linked Hormone Receptor

cyclic AMP. Cyclic AMP is often termed the "second messenger," because it delivers the message inside the cell, after the first message—the hormone—is received at the surface of the cell. For other hormones, the G protein stimulates a different cascade of proteins, which ultimately use calcium levels instead of cylic AMP to distribute the internal signal. Activation of the G protein by a hormone then leads to release of the calcium into the cytoplasm, where it is sensed by *calmodulin*. Because of its central role in the sensing of signals, calmodulin is a very plentiful protein, often comprising up to 1% of the total cellular protein.

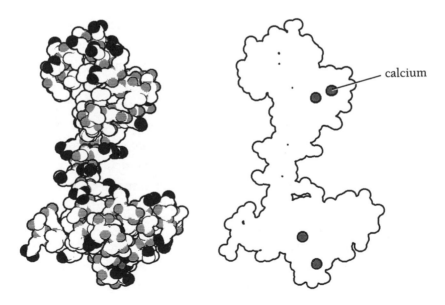

Calmodulin

Sense

Our five senses link us to the exterior world. We test food for palatability by taste and smell. We locate other people, friends and foes, and are warned of impending dangers by the sounds they make. By touch, we explore the intimate world of surface textures, heat, and cold. And by sight, we recognize millions of different objects. Without these five windows onto the world, we would be locked inside our individual interior worlds, lost in circles of thought.

The senses translate physical properties—like light and sound—into messages the brain can understand. The tongue and nose contain cells that search out different small molecules, classify them as being salty or bitter, fragrant or foul, and dispatch a nerve signal with the answer. Cells in the ears monitor tiny hairs for motion. If sound waves set them vibrating, they send a nerve message to the brain describing the pitch and loudness. Cells in the skin monitor heat, cold, and pressure, warning the brain if any approach dangerous levels. The eyes act as small cameras, focusing an image of light onto the retinal cells, which send messages to the brain describing the color and brightness of their particular view.

Opsins

Light is sensed through a small change in the shape of *retinal*. Normally in a kinked shape, retinal abruptly straightens when it absorbs a photon of light. Retinal is a small, carbon-rich molecule made from vitamin A. We cannot make retinal from scratch and must obtain the starting materials in the diet. Liver is a rich source of vitamin A, and is a traditional folk cure for night blindness. *Carotene*, found in carrots and dark green vegetables, may be broken in half to yield retinal, so these vegetables are also cures for weak vision. Opsin, a membrane-bound protein, surrounds retinal, feeling this change of shape. When it feels a change, opsin launches a cascade similar to that of the G-linked hormone receptors. The end result of the cascade is not the activation of certain enzymes, as in the hormone cascade, but instead the launching of a nerve signal to the visual processing centers of the brain.

Color vision is the result of differences in the opsin proteins holding each molecule of retinal. Night vision is due to *rhodopsin*, which senses greenish-blue light best. Cells carrying rhodopsin are so sensitive that a single photon of light may be detected. Rhodopsin gives no color information, so the night world is a world of black and white (or more properly, black and blue-green). Color is sensed by three different opsins: one most sensitive to red light, one to green light, and one to blue light. Each individual cell in the retina contains only one of these three opsins, and thus responds to only one color. Surprisingly, the three types are not present in equal proportions: for every 20 red-sensitive cells, there are 40 green-sensitive cells and only one blue-sensitive

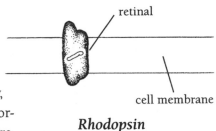

Rhodopsin

cell. This makes our 'blue' image of the world grainier than our red and green views.

Color blindness occurs when one or more of these opsins are congenitally absent. The most common is red/green color blindness, caused when either the red or the green opsin is missing, which occurs in about one of every 12 men and one of every 200 women. The higher incidence in men is due to the fact that the instructions for building the red and green opsins are located in the X chromosome, so the color blindness is genetically *sex-linked*. Women carry two X chromosomes in each cell, but men carry only one, inherited from their mothers (men carry a Y chromosome, inherited from their fathers, in place of the second X chromosome). Since men only have one X chromosome, they only have one copy of the opsin instructions. If these instructions are faulty, the result is color blindness. But women have two copies of the opsin instructions; if one is faulty, the other will serve. Only when both are faulty, when both parents provide a faulty gene, does red/green color blindness appear in women.

A common misconception has its basis in the red, green, and blue opsins of color vision. We often hear that mixing red light and green light makes yellow light. In reality, we merely *see* red+green light as being yellow. This mixture of two colors, a muddle of wavelengths, is sensed similarly to pure yellow light, taken from the middle of the yellow part of the spectrum. Both the mixture and the pure light stimulate the red and the green opsins, and both are interpreted as "yellow." Because of this limitation in color vision, color televisions fool our eyes into seeing "full color" by using only three colors: red, green, and blue—one for each of our opsins.

Crystallins

Crystallins belie their name—they are proteins that are designed *not* to form crystals. In our eyes, a concentrated solution of proteins is used as a lens to focus

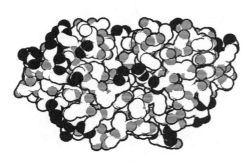
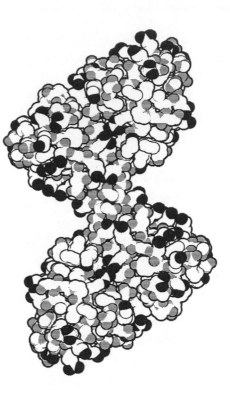

Crystallins

light, just as glass is used in camera lenses. Eye lenses are composed of cells filled with crystallin proteins, comprising one-third to one-half of the weight of the cell. At these high concentrations, most proteins would aggregate, forming particles or tiny crystals that would scatter light, making the lens opaque. To solve this problem, the lenses contain several different crystallins, with different sizes and shapes. Mixed together, they do not form crystals or complexes, but stay in a glassy random solution.

Eye lens cells, during early development, fill themselves with crystallins, allowing the machinery of the cell—nucleus, mitochondria, ribosomes—to atrophy as the concentration mounts. In this way, they are similar to red blood cells. Red blood cells are containers filled with hemoglobin, lens cells are containers filled with crystallins. Because the nucleus and all of the protein synthesis machinery have been discarded, eye lenses cannot build new crystallins when they wear out. Crystallins belong to a rare class of proteins that are made for life. Unfortunately, they often do not last long enough. With age, they tend to aggregate and form sulfur–sulfur linkages from protein to protein. These aggregates cloud the lens, forming *cataracts*. Today, the clouded lens may be removed and replaced by a plastic lens. Surprisingly, this may allow the patient to see ultraviolet light, which is normally absorbed and filtered out by the crystallins.

Odorant Receptors

We smell fragrant molecules with odorant receptors in the nose, similar in structure to the opsin proteins of vision. But instead of watching a light-sensitive molecule for change, odorant receptors monitor a smell-sensitive binding pocket, tailored to fit one particular molecular shape. When a molecule binds in its pocket, a receptor "smells" the molecule and sends a message to the brain. An odorant receptor with a small pocket binds to small neutral molecules, sending a message of "gasoline." A larger pocket fits large, flat neutral molecules, like naphthalene, smelling "mothballs." Small esters—molecules composed of an organic acid connected to an alcohol—fit into pockets lined with oxygen- and nitrogen-rich amino acids, sending the message "banana" or "wintergreen." Dozens of these receptors have been explored in the laboratory: smoky, flowery, sweaty, fishy, musky, minty, malty. It has been estimated that we make several hundred different receptors, each waiting for one specific molecular shape—one particular smell.

The sense of smell is quite variable from person to person. Just as a genetic absence of one opsin leads to color blindness, genetic deficiency of one of the odorant receptors will yield a "blindness" for a given smell. Smell blindness is thought to be a recessive trait, just like blue eyes: if both parents cannot smell camphor, their child will not either. Smell blindness for a given smell is highly variable. Most common smells, such as those that allow us to differentiate spoiled food from fresh, are detected by nearly everyone. But other specialized smells may only be detected by some. The molecule androstenone is used for research in smell blindness. It is a steroid (page 149) that is isolated from pigs and that surprisingly is also present in celery and truffles. To those who can smell it, it has a strong scent, but almost one half of the population cannot.

Taste Receptors

Much of the pleasure we experience while eating food—aromatic wines, smoky seared meats, exotically flavored tropical fruits—is actually a consequence of their odors. The multitude of different odorant receptors lining the nose lend these delicate shades to food. Taste, by contrast, is starkly simple: there are four basic tastes—sour and salty, bitter and sweet—and perhaps one more—the meaty taste of amino acids. Cells in the taste buds, dotting the surface of the tongue, are sensitive to each of these tastes.

Bitter and sweet foods are sensed like hormones and odors. A specific receptor protein looks for these substances—sugars in the case of sweet foods and various poisonous molecules for bitter foods—and dispatches a message to the brain. Sweet tastes are pleasant because they signal useful sugars in the diet, which may be used directly for energy. Bitter tastes are unpleasant because they signal dangerous compounds, like toxic alkaloids found in poisonous plants. Quinine, strychnine, nicotine, and caffeine all yield bitter tastes. Taste receptors can be fooled. Artificial sweeteners like saccharine (675 times sweeter than table sugar) and aspartame (150 times sweeter than table sugar) fool the sweet receptors, promising nourishment, but providing only indigestible, zero-calorie compounds.

Sour and salty foods are sensed more simply. Sodium ions taste salty; potassium ions taste bitter and slightly less salty. As discussed below, the transport of sodium across a membrane is the basis of nerve signals. So, cells sensing salty tastes need merely to build a channel for sodium ions, allowing them to enter when they are present in the food. The change in sodium level itself launches the signal. Sour tastes reveal foods that are acidic. The acetic acid in vinegar and the citric acid in orange juice and sour confections are two common culinary examples. The hydrochloric acid in digestive fluids, a much stronger acid, would taste twice as sour as these two. The hydrogen ions in acidic foods affect the same sodium channels used for sensing salty foods, by blocking their action. This numbing of salty receptors is interpreted as a sour taste. Foods that contain large amounts of both salt and acid may have only a faint metallic taste, because of the antagonistic actions of sodium ions and acidity on these channels.

A separate gated channel opens when it binds to amino acids, allowing ions to enter and initiating a nerve signal with the message "meaty." The taste of amino acids is separate from the four tastes commonly described in Western literature and is given the name *umami* in Japanese. Monosodium glutamate (MSG), a salt of the amino acid glutamate, is the most common spice for adding this flavor. It activates these channels and is interpreted as a meaty taste. MSG is commonly used as a flavor enhancer. Like sweet and salty tastes, its pleasant taste signals a food of nutritive value.

Thought

Our thoughts are the culmination of a continuous ebb and flow of sodium ions. Nerves use changing levels of sodium to transmit messages within the brain and out to distant parts of the body. Nerve signals are transmitted over enormous distances (our arms are enormous to an individual cell) along the extended *axons* of nerve cells. Axons are energized, making them ready to deliver a signal, by depleting their supply of sodium. Then, in a wave that spreads down the axon, the sodium floods back in and a message flashes down the axon. Some nerve signals are quite direct. When we decide to close our hand, the brain dispatches a signal down a long axon, straight to the muscle. There, *neurotransmitters* are released from the tip of the nerve. Crossing the tiny gap to the muscle cells, termed a *synapse*, they stimulate the muscle to contract. These messages flash down nerve fibers at up to 100 meters per second, and nerve cells recover so quickly that 200 messages may be sent every second.

Nerve signals may also be marvelously complex. In the brain, a vast network of connecting axons, combined with precise weighing of signals by each nerve cell, creates our interior world of thought. Current ideas of brain function suggest that thoughts—memories, ideas, inspirations—are formed by the combined actions of many nerve cells. There is not one particular cell that, when excited, recalls the memory of that tree you climbed as a child or the smell of library paste. Instead, the memory is a specific play of signals through a network of carefully connected nerve cells.

Sodium—Potassium Pump

The sodium–potassium pump depletes nerves of sodium, priming them for a nerve impulse. Embedded in the membrane, they continually transport sodium ions out of axons. This creates an electrical difference across the membrane, charging the axon like a battery. The sodium–potassium pump uses ATP to force sodium against its natural tendency to flow back inside. For each ATP consumed, three sodium ions are transported out and two potassium ions are simultaneously transported inwards. These sodium ions, when given the chance, rush back into the cell, attempting to equalize the amounts inside and outside. In nerve cells, the vigor of this flow is used to fuel the transmission of messages. In most of our other cells, a similar sodium gradient powers other cellular processes. Proteins embedded in the cell membrane fuel the import of sugars, amino acids, and other nutrients into the cell by allowing a few sodium ions to leak in at the same time.

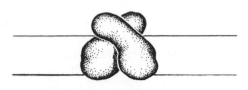

Sodium–Potassium Pump

Other proteins exchange a hydrogen ion for a sodium ion—hydrogen out, sodium in—allowing cells to control their internal acidity. These processes are so important that about one-third of the ATP that we make goes to fueling this pump. In nerves, it may consume up to two-thirds of the available energy.

Digitoxin and *digoxin*, poisons isolated from two varieties of foxglove, attack the sodium–potassium pump. In trace amounts, these poisons are used therapeutically. Small doses gently slow the transport of sodium out of cells, allowing the level of sodium inside to mount. When the level gets too high, the cells panic and call a different pump into action. It transports sodium out by allowing calcium, instead of potassium, to enter at the same time. Calcium then slowly builds up in the cell as the normal low level of sodium is restored. The elevated level of calcium strengthens the force of muscle contraction, which may be beneficial for people with weak hearts. But even a moderate overdose of digitoxin—perhaps three or four times this low prescribed amount—is deadly. The level of calcium in the muscle cells skyrockets, preventing them from relaxing and leading to heart attack.

Voltage-Gated Ion Channels

Messages are relayed down the long axons of nerves, often over distances thousands of times longer than a typical cell, by voltage-gated ion channels. Under normal conditions—with the membrane fully charged, rich in sodium outside and depleted of sodium inside—these channels remain tightly closed. But as the sodium levels equalize, reducing the voltage across the membrane, the channels snap open, allowing the sodium ions in the local vicinity to flow inside. To send a signal down a charged axon, the nerve trips the first few channels. They deplete the sodium charge in the local area. This trips neighboring channels further down the axon, opening them, and depleting their local environment. In a chain reaction, a wave of opening channels sweeps down the axon, carrying the message. The open channels are unstable and close spontaneously in a thousandth of a second. The sodium–potassium pump then works to restore the sodium charge, priming the axon for the next signal.

Extract of monkshood—aconite—is a classical poison of mythology and history. The active molecule is *aconitine*, an alkaloid that impairs the action of voltage-gated ion channels. Aconitine forces the channels into the open form, completely depleting the imbalance of sodium. As the poison spreads, the victim dies as all communication ceases. Obviously, this channel is a vital link, making it an attractive target for plants and animals that wish to protect themselves. Many other organisms make toxins that poison voltage-gated ion channels, either by forcing all of the channels open or by locking them tightly shut. Poisonous newts and poison-dart frogs, deadly cone shells, rhododendron flowers, sea anemones, the dinoflagellates causing red tide, and scorpions (page 112) all make toxins that attack these channels.

Neurotransmitters

At the end of a nerve axon, signals are relayed to the next cell across the tiny gap of a synapse. In some cases, the same sodium-based signal that sweeps down the axon is passed directly to the next nerve cell. In most cases, however, the signal is converted into the chemical signal of neurotransmitters. The tip of the axon holds many small containers filled with neurotransmitters. When the sodium signal reaches the tip, it releases these neurotransmitters into the gap. They diffuse across and are picked up by receptor proteins, like the acetylcholine receptor (see below), on the neighboring nerve cell. Neurotransmitters are very small molecules,

Neurotransmitters

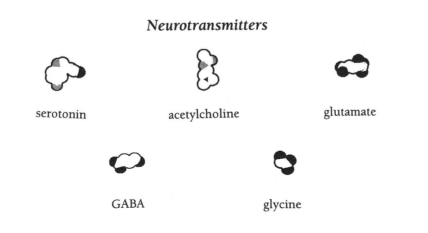

serotonin   acetylcholine   glutamate

GABA   glycine

so that they may travel quickly through the crowded synapse. Two of those pictured here are particularly economical to make: glycine and glutamate are amino acids naturally found in proteins.

Neurotransmitters have a variety of effects. Some excite the neighboring cell, causing it to send a signal down its axon. Others inhibit the cell, telling it to be cautious about sending a signal. Any particular nerve may have thousands of synapses peppering its surface with neurotransmitter messages, delivering both excitatory and inhibitory signals. By weighing the relative strengths of all of these signals, excitatory against inhibitory, the cell determines when a signal should be dispatched down its own axon. This allows remarkable control in the brain. Nerve signals, sweeping down an axon, are either "on" or "off," but the messages transmitted by neurotransmitters may be shaded at infinite levels between.

The opposite actions of inhibitory synapses and excitatory synapses are easily seen by the different effects of drugs that block their action. Depressants such as *alcohol*, *barbiturates*, and *diazepam* (Valium) enhance the binding of inhibitory neurotransmitters to their receptors. In small amounts, this reduces the inhibitions of the consumer, slowing the flashing warnings traveling throughout the nervous system. In excessive amounts, or when consumed together, this inhibition may reach deadly levels, as all thoughts grind to a slow halt. *Strychnine* has the opposite effect, blocking the binding of inhibitory neurotransmitters to receptors. Without the controlling effects of these inhibitory synapses, nerve signals run rampant throughout the body. In the throes of strychnine poisoning, any small stimulus—a flash of light, a touch, a loud sound—causes a flurry of nerve impulses. These ultimately reach muscles cells, stimulating a massive contraction of muscles throughout the body, causing the painful spasms of strychnine poisoning.

Amphetamines enhance the action of excitatory synapses. They are similar to the neurotransmitters *norepinephrine* and *dopamine*, fooling the synaptic machinery that uses them. Amphetamines elevate nervous activity, reducing fatigue and sharpening attention. Psychedelic drugs, such as *LSD* (lysergic acid diethylamide) and *mescaline*, enhance the action of a small class of excitatory receptors that use the neurotransmitter *serotonin*. These synapses are intimately involved in processing sensory information, so the confusion caused by these drugs can lead to sensory hallucinations—phantoms created inside the brain. Poisons that block excitatory receptors, such *curare* and *hemlock*, have a much more direct effect. They block the excitatory signals from the brain to muscles, causing instant paralysis.

Acetylcholine Receptor

When we decide to move an arm or flex a foot, the brain dispatches a signal down a long motor neuron. When the signal reaches the end, the neurotransmitter *acetylcholine* is released. It crosses the narrow synapse and binds to acetylcholine receptors on the target muscle cell. The receptors snap open and sodium ions flood into the cell, initiating a cascade of actions that ultimately causes the muscle cell to contract. Soon after, the excess acetylcholine, still floating in the synapse, is destroyed by the enzyme *acetylcholine esterase*. This cleanup operation is important: if the excess neurotransmitter were allowed to remain in the gap, it would continuously stimulate the muscle cell, never allowing it to relax.

The link from brain to muscle is critical. Without it, we cannot move. The acetylcholine receptor is one of our many molecular Achilles' heels, an easy target for poisons and toxins. Plants, in particular, create poisons that attack the acetylcholine receptor. Since they have no nerves themselves, plants may produce neuromuscular poisons—such as *curare, atropine,* and *hemlock*—with no risk to themselves. In trace amounts, these poisons have found medical application: curare is used to relax muscles during surgery, and atropine is used in eye drops to dilate the pupils during an eye exam, temporarily paralyzing the muscles in the irises. But in larger amounts, such as that from a few leaves of hemlock, these poisons cause total paralysis.

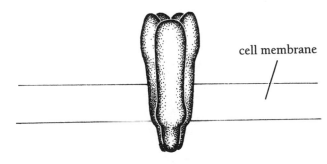

cell membrane

Acetylcholine Receptor

Many synapses do not merely transmit messages. Some also have the remarkable ability to learn. If an individual synapse is heavily used, it may be made more sensitive to the next message, oiling well-used nerve pathways in the brain. The NMDA receptor (NMDA for *N*-methyl-D-aspartate, a molecule that has been used in its study) is an integral part of this process. If glutamate is applied to this receptor when messages are flashing though a nerve cell, the synapse is made more sensitive. This is thought to be the molecular basis of *memory*. These sensitized synapses "remember" their messages, and together, a network of sensitized nerve cells "remember" a thought, a sound, or an image.

Epilogue:
Molecules and Medicine

In general, our molecules perform their tiny duties without any conscious help. Food is digested, transported, and broken down for energy automatically. Myosin and actin seamlessly synchronize their motion to contract and release our muscles on demand. Proteins are made when needed and destroyed when worn. We merely have to provide enough food, a few essential vitamins, enough air and enough water, and stay away from extremes of heat and cold.

Occasionally, our molecules do not behave as we would wish. Histamine may be formed when not absolutely necessary, causing allergies to fairly harmless substances, such as pollen or dust. Pain may stab far out of proportion to the danger: headaches may warn of tension or fatigue instead of a life-threatening peril. Excess stomach acid may cause extreme, but far from lethal, discomfort. Centuries ago, a person had little recourse but to weather the problem. But today we routinely interact with our own molecules, modifying their action to suit our needs. We take antihistamines, which circulate through the body and block the proteins that make and use histamine. The symptoms of allergy cease. We take analgesics that seek out the molecules that transmit pain, so the signal never reaches the brain. Pain dulls to acceptable levels. We swallow chalky salts or take molecules that block the enzymes that make stomach acid, neutralizing the contents of the stomach. Digestion may be slightly impaired, but the heartburn subsides. Hundreds of products are available for personally modifying the action of our own molecules.

Modern medicine pulls no punches toward the molecules of infecting organisms. Every possible method of blocking, removing, or destroying them has been explored. Antiseptics poison bacteria on the skin or in an injury. Antibiotics flow through the body, destroying bacterial cell walls, poisoning their energetic machinery, or starving them to death. Antiviral drugs neutralize the infectious coatings of viruses, or block their reproduction. Vaccinations strengthen our natural defenses, priming us for defense. Often, modern medicine also takes a cavalier approach to our own molecules. Natural poisons may be used temporarily to paralyze or relax muscles during surgery or during an examination. A diverse collection of drugs may be administered to gently shade the action of nerve transmission, lifting depression, easing stress, or sedating paranoia. Consciousness may be anesthetized to shield a patient from the traumatic experience of surgery. Thousands of drugs, natural and synthetic, are available to the physician, and hundreds more are developed each year to fight new molecular battles.

With knowledge comes control. As our understanding of the intimate details of each of our molecules grows, so does our ability to modify, carefully, their delicate actions and interactions. Our control may as simple as curbing the amount of fat in the diet to reduce the risk of heart disease, or as extreme as infecting every cell in the body with a foreign gene to reverse a genetic deficiency. Never before have the possibilities for betterment of life been so profound.

Further Reading

Alberts B, Bray D, Lewis J, Raff M, Roberts K & Watson JD. *Molecular Biology of the Cell*. 3rd ed. New York: Garland Publishing, 1994.

Atkins PW. *Molecules*. New York: Scientific American, 1987.

Branden C & Tooze J. *Introduction to Protein Structure*. New York: Garland Publishing, 1991.

de Duve C. *A Guided Tour of the Living Cell*. New York: Scientific American, 1984.

Devlin TM, ed. *Textbook of Biochemistry with Clinical Correlations*. New York: Wiley-Liss, 1992.

Dickerson RE & Geis I. *The Structure and Action of Proteins*. New York: Harper and Row, 1969.

Goodsell DS. *The Machinery of Life*. New York: Springer-Verlag, 1993.

Stevens SD & Klarner A. *Deadly Doses: A Writer's Guide to Poisons*. Cincinnati: Writer's Digest, 1990.

Stryer L. *Biochemistry*. 4th ed. New York: WH Freeman, 1995.

Glossary

active site The portion of an enzyme directly involved in catalysis of a chemical reaction, often contained in a shallow depression or deep cleft on the enzyme surface.

allosteric enzyme An enzyme that undergoes a change in shape that modifies its function.

amino group A nitrogen atom carrying two hydrogen atoms: $-NH_2$.

amino acid The building blocks of proteins: small molecules containing an amino group and an acid group. Twenty chemically diverse amino acids are commonly linked together to form proteins.

antibiotic A substance, such as penicillin, that inhibits the growth of bacteria or other microorganisms.

atom The smallest part of an element that retains its individual properties. Human molecules are composed primarily of carbon, hydrogen, oxygen, and nitrogen atoms. Phosphorous, sulfur, and various metal atoms are found in smaller amounts.

axon A long, narrow appendage of a nerve cell, along which nerve signals are transmitted.

biotechnology The use of engineering to modify the biological characteristics of living organisms.

carbohydrate A diverse group of molecules containing carbon, hydrogen, and oxygen atoms in the ratio 1:2:1, including simple sugars, such as glucose and fructose, and polysaccharides, such as starch.

cascade A series of enzymes, each of which activates an increasingly large number of the next, amplifying a signal. Examples include the complement cascade and the blood clotting factors.

catalyst A substance or molecule, such as an enzyme, that facilitates a chemical reaction without itself being consumed in the process.

cell   The smallest part of a living organism that is capable of performing all of the activities of life.

chemotherapy   Treatment with chemicals that are toxic to disease-producing micro-organisms or cancerous cells.

chromosome   Small, threadlike bodies found in the cell nucleus, each containing one strand of DNA surrounded by protective proteins.

codon   Three successive nucleotides in a DNA or RNA strand, used to specify one amino acid in a protein chain.

cytoplasm   The contents of a cell bounded by the cell membrane, excluding the nucleus.

cytoskeleton   A network of protein filaments that extend throughout the cytoplasm, supporting the cell and providing a structure for transport.

disaccharide   A "double sugar," composed of two simple sugars chemically linked together. Examples include sucrose, composed of glucose and fructose, and lactose, composed of glucose and galactose.

digestion   The process of breaking food into molecules small enough to be absorbed.

electron   An elementary particle with a negative charge. The outermost electrons in atoms are shared in chemical bonding, or may be removed entirely to form ions.

enzymes   A class of proteins that catalyze chemical reactions, most often given names that end with the suffix -ase.

evolution   The gradual change of life over the history of the Earth, due to selection of favorable genetic mutations.

gene   A discrete unit of hereditary information encoded in a DNA strand.

hormone   A molecule, such as insulin or growth hormone, used to deliver signals through the blood.

hydroxyl group   An oxygen atom carrying a single hydrogen atom: --OH.

ion   An electrically-charged atom, positively-charged if it has lost electrons or negatively-charged if it has gained electrons.

lipid   A family of carbon-rich molecules that are insoluble in water, including fats, phospholipids, and cholesterol.

lysosome   A membrane-enclosed compartment in the cell cytoplasm, containing digestive enzymes.

membrane   A thin, relatively impermeable sheet, composed of a lipid bilayer, used to surround cells or compartments within cells.

methyl group   A carbon atom carrying three hydrogen atoms: $-CH_3$.

mitochondrion   A compartment in the cell cytoplasm surrounded by two membranes, containing most of the proteins used to generate metabolic energy.

molecule   A collection of atoms held together by chemical bonds.

mutation   A change in the sequence of nucleotides in DNA. The term also refers to changes in the organism caused by this modified DNA.

neurotransmitter   A small molecule used to transmit nerve messages from cell to cell across a synapse.

nucleic acid   Large, linear molecules composed of a chain of nucleotides: DNA is used to store the genetic information, RNA utilizes this information to build proteins.

nucleotide   The building blocks of nucleic acids, composed of a sugar, a phosphate, and a base group.

nucleus   The large compartment at the center of cells, surrounded by a double membrane and containing the cellular DNA.

peptide   A short protein chain composed of less than a few dozen amino acids.

phosphate   A phosphorous atom carrying four oxygen atoms. May be found as a free ion or connected to a molecule in the form of phosphate groups, as in ATP.

protein   Large molecules composed of chains of amino acids, most often folded into a compact globule.

subunit   Individual protein chains in a protein complex composed of many chains.

synapse   The narrow gap between cells in the nervous system.

vitamin   Small molecules required in the diet, necessary for key metabolic functions.

Sources of
Macromolecular Structures

Accession codes, such as PDB4CHA for chymotrypsin, are given for coordinates taken from the Brookhaven Protein Data Bank. These coordinates are available on the World Wide Web at `http://www.pdb.bnl.gov/`.

Chapter 2

Chymotrypsin (cow): Tsukada & Blow (1984) PDB4CHA
Carbonic Anhydrase (human): Kannan, Ramanadham & Jones (1983) PDB2CAB
Aspartate Carbamoyltransferase (bacteria): Stevens, Gouaux & Lipscomb (1990) PDB4AT1
Tryptophan Synthase (bacteria): Hyde, Ahmed, Padlan, Miles & Davies (1988) PDB1WSY
Aspartate Aminotransferase (chicken): McPhalen, Vincent & Jansonius (1991) PDB7AAT
Dihydrofolate Reductase (human): Davies & Kraut (1989) PDB1DHF
Thymidylate Synthase (bacteria): Montfort & Stroud (1991) PDB2TSC
Glutamine Synthase (bacteria): Eisenberg, Almassy & Yamashita (1989) PDB2GLS
Alcohol Dehydrogenase (horse): Al-Karadaghi & Cedergren-Zeppesauer (1993) PDB2OHX
D-Xylose Isomerase (fungal): Mrabet et al. (1991) PDB1XIM
DNA: idealized B-helix
p53 Tumor Suppressor (human): Cho, Gorina, Jeffrey & Pavletich (1994) Science 265, 346
Nucleosome (chicken): Arents, Burlingame, Wang, Love & Moudrianakis (1991) Proc. Natl. Acad. Sci. USA 88, 10148
DNA Topoisomerase I (bacteria): Lima, Wang & Mondragon (1994) Nature 367, 138
DNA Gyrase (bacteria): Kirchhausen, Wang & Harrison (1985) Cell 41, 933
DNA polymerase (bacteria): Beese, Friedman & Steitz (1993) PBD1KFD
Messenger RNA: idealized model
RNA Polymerase I (yeast): Schultz, Celia, Riva, Sentenac & Oudet (1993) EMBO J. 12, 2601
RNA Polymerase II (yeast): Darst, Edwards, Kubalek & Kornberg (1991) Cell 66, 121
met Repressor (bacteria): Somers & Phillips (1992) PDB1CMA
Catabolite Gene Activator Protein (bacteria): Schultz, Shields & Steitz (1991) PDB1CGP
Zif268 (mouse): Pavletich & Pabo (1992) PDB1ZAA
GCN4 (yeast): Ellenberger & Harrison (1993) PDB1YSA
Transfer RNA (yeast): Jack, Ladner & Klug (1978) PDB4TNA
Seryl-tRNA Synthetase (bacteria): Fujinaga, Berthet-Colominas & Cusack (1993) PDB1SRY

Glutaminyl-tRNA Synthetase (bacteria): Rould, Perona, Soll & Steitz (1989) Science 246, 1135
Ribosome mitochodrial 70S (bacteria): Frank, Penczec, Grassucci & Srinvastava (1991) J. Cell
    Biol. 115, 597
Ribosome cytoplasmic 80S (rabbit): Verschoor & Frank (1990) J. Mol. Biol. 214, 737
GroEL (bacteria): Braig, Otwinowski, Hegde, Boisvert, Joachimiak, Horwich & Sigler (1994) Na-
    ture 371, 578
Proline *cis-trans* Isomerase (human): Ke (1992) PDB2CPL
Disulfide Bond Formation Protein (bacteria): Martin, Bardwell & Kuriyan (1993) PDB1DSB
Ubiquitin (human): Cook, Jeffrey, Kasperek & Pickart (1993) PDB1TBE
Proteosome (cow): Gray, Slaughter & DeMartino (1994) J. Mol. Biol. 236, 7

Chapter 3

Trypsin (cow): Huber & Deisenhofer (1982) PDB2PTC
Chymotrypsin: Chapter 2
Elastase (pig): Takahashi, Radhakrishnan, Rosenfield, Meyer & Trainor (1989) PDB4EST
Trypsin inhibitor (cow): Huber & Deisenhofer (1982) PDB2PTC
Pepsin (pig): Cooper, Khan, Taylor, Tickle & Blundell (1990) PDB5PEP
Chymosin B (cow): Newman, Frazao, Khan, Tickle, Blundell, Safro, Andreeva & Zdanov (1991)
    PDB4CMS
Cathepsin D (human): Baldwin, Bhat, Gulnik & Erickson (1993) PDB1LYB
Carboxypeptidase A (cow): Lipscomb & Rees (1982) PDB4CPA
Aminopeptidase (pig): Hussain, Tranum-Jensen, Noren, Sjostrom & Christiansen (1981) Biochem.
    J. 199, 179
Dipeptidyl Peptidase IV (pig): Hussain (1986) Biochem. Biophys. Acta 815, 306
Amylase (fungus): Swift, Brady & Derewenda (1992) PDB6TAA
Glucoamylase (fungus): Harris, Aleshin, Firsov & Honzatko (1993) PDB1DOG
Disaccharidases (rabbit): Noren, Sjostrom, Danielsen, Cowell & Skovbjerg (1986) Molecular and
    Cellular Basis of Digestion, Elsevier, Amsterdam
Phospholipase A_2 (human): Scott, White & Sigler (1992) PDB1POE
LDL (human): Antwerpen & Gilkey (1994) J. Lipid Res. 35, 2223
HDL (human): Ohtsuki, Edelstein, Sogard & Scanu (1977) Proc. Natl. Acad. Sci. USA 74, 5001
Ribonuclease A (cow): Berdsall & McPherson (1992) PDB1RTA
Deoxyribonuclease I (cow): Lahm & Suck (1986) PDB2DNJ
ATP: idealized coordinates
Hexokinase (yeast): Bennett & Steitz (1980) PDB1HKG
Glucose-6-Phosphate Isomerase (pig): Muirhead (1977) PDB1PGI
Phosphofructokinase (bacteria): Shirakihara & Evans (1988) B1PFK
Aldolase (human): Watson (1991) PDB1ALD
Triose Phosphate Isomerase (yeast): Davenport, Bash, Seaton, Karplus, Petsko & Ringe (1991)
    PDB7TIM
Glyceraldehyde-3-Phosphate Dehydrogenase (human): Watson & Campbell (1983) PDB3GPD
Phosphoglycerate Kinase (yeast): Shaw, Walker & Watson (1982) PDB3PGK
Phosphoglycerate Mutase (yeast): Campbell, Hodgson, Warwicker, Winn & Watson (1982)
    PDB3PGM
Enolase (yeast): Lebioda & Stec (1990) PDB5ENL
Pyruvate Kinase (cat): Muirhead, Levine, Stammers & Stuart (1980) PDB1PYK
Pyruvate Dehydrogenase Complex (mammal): Reed & Hackert (1990) J. Biol. Chem. 265, 8971
Citrate Synthase (pig): Remington, Wiegand & Huber (1984) PDB1CTS
Aconitase (pig): Robbins & Stout (1990) PDB6ACN
Isocitrate Dehydrogenase (bacteria): Hurley, Dean, Koshland & Stroud (1991) PDB9ICD
α-Ketoglutarate dehydrogenase complex (mammal): Reed & Hackert (1990) J. Biol. Chem. 265,
    8971; modeled after Mattevi, Obmolova, Schultz, Kalk, Westphal, DeKok & Hol (1992) Sci-
    ence 255, 1544

Succinyl-CoA Synthetase (mammal): Wolodko, Fraser, James & Bridger (1994) J. Biol. Chem. 269, 10883
Succinate Dehydrogenase (cow): Ohnishi (1987) Curr. Topics Bioenerg. 15, 37
Fumarase (pig): Sacchettini, Meininger, Roderick & Banaszak (1986) J. Biol. Chem. 261, 15183
Malate Dehydrogenase (pig): Birktoft & Banaszak (1989) PDB4MDH
NADH Dehydrogenase (fungus): Hofhaus, Weiss & Leonard (1991) J. Mol. Biol. 221, 1027
Coenzyme Q: idealized coordinates
Cytochrome Reductase (fungus): Weiss, Leonard & Neupert (1990) TIBS 15, 178
Cytochrome c (tuna): Tanako (1980) PDB3CYT
Cytochrome Oxidase (bacteria): Iwata, Osterheimer, Kudwig & Michel (1995) Nature 376, 660
ATP Synthase (cow): Abrahams, Leslie, Lutter & Walker (1994) Nature 370, 621
Hemoglobin (human): Shaanan (1983) PDB1HHO
Myoglobin (sperm whale): Phillips (1981) PDB1MBD
Ferritin (horse): Precigoux, Yariv, Gallois, Dautant & Courseille (1993) PDB1HRS
Transferrin (human): Baker, Anderson & Haridas (1992) PDB1LFG
Adenylate Kinase (pig): Schulz (1987) PDB3ADK
Ncleoside Diphosphate Kinase (fungus): Janin, Morera, Dumas, Lascu, Lebras & Veron (1993) PDB1NDP

Chapter 4

Phospholipid and Cholesterol: idealized coordinates
Porin (bacteria): Weiss & Schultz (1992) PDB2POR
Connexon (rat): Unwin & Ennis (1984) Nature 307, 609
Nuclear Pore (frog): Hinshaw, Carragher & Milligan (1992) Cell 69, 1133
Calcium Pump (sea scallop): Toyoshima, Sasabe & Stokes (1992) Nature 362, 469
Actin (rabbit): Kabsch, Mannherz, Suck, Rai & Holmes (1991) PDB1ATN
Microtubules (various): Amos & Eagles (1987) in Fibrous Protein Structure (Squire & Vibert, eds) Academic Press, London, p 215
Keratin (human): Fuchs, Tyner, Giudice, Marcuk, Ray, Chaudhury & Rosenberg (1987) Curr. Topics Dev. Biol. 22, 5
Collagen (human): Nementhy, Gibson, Palmer, Yoon, Paterlini, Zagari, Rumsey & Scheraga (1992) PDB1BBE, van der Rest & Garrone (1991) FASEB J. 5, 2814
Hyaluronic Acid: idealized model
Proteoglycan (mammal): in (1986) Functions of the Proteoglycans, Ciba Foundation Symposium 124, John Wiley & Sons, New York
Elastin (cow): Raju & Anwar (1907) J. Biol. Chem. 262, 5755
Fibronectin (human): Engel, Odermatt, Engle, Madri, Furthmayr, Rohde & Timpl (1981) J. Mol. Biol. 150, 97
Tenascin (human): Erickson & Inglesias (1984) Nature 311, 267
Laminin (mouse): Engel, Odermatt, Engle, Madri, Furthmayr, Rohde & Timpl (1981) J. Mol. Biol. 150, 97
Myosin (chicken): Rayment & Holden (1993) PDB1MYS
Kinesin (sea urchin): Scholey, Heuser, Yang & Goldstein (1989) Nature 338, 355
Dynein (algae): Goodenough & Heuser (1984) J. Mol. Biol. 180, 1083

Chapter 5

Cobra Neurotoxin: Betzel, Lange, Pal, Wilson, Maelicke & Saenger (1991) PDB2CTX
Cobra Cardiotoxin: Rees, Bilwes, Samama & Moras (1990) PDB1CDT
Scorpion Neurotoxins: Zhao, Carson, Ealick & Bugg (1992) PDB2SN3, Bontems, Roumestand, Gilquin, Menez & Toma (1993) PDB2CRD
Melittin (honeybee): Eisenberg, Gribskov & Terwilliger (1990) PDB2MLT
Enterotoxin (bacteria): Sixma & Hol (1992) PDB1LTS
Cytochrome p450cam (bacteria): Poulos (1992) PDB1PHA

Glutathione: idealized coordinates
Glutathione S-Transferase (human): Huber & Parker (1992) PDB1GSS
Superoxide Dismutase (cow): Tainer, Getzoff, Richardson & Richardson (1980) PDB2SOD
Catalase (cow): Murthy, Reid, Sicagnano, Tanaka, Fita & Rosmann (1984) PDB7CAT
Lysozyme (human): Artymiuk & Blake (1984) PDB1LZ1
α-Macroglobulin (human): Sottrup-Jensen (1989) J. Biol. Chem. 264, 11539
Metallothionein (rat): Robbins & Stout (1993) PDB4MT2
Antibody (human): Diesenhofer (1981) PDB1FC1, Marquart & Huber (1989) PDB2FB4, Klein (1990) Immunology, Blackwell Scientific Publications, Boston
MHC (human): Madden, Gorga, Strominger & Wiley (1992) PDB1HSA
CD4 (human+rat): Brady, Dodson & Lange (1993) PDB1CID, Wang, Yan, Garrett & Harrison (1990) PDB2CD4
CD8 (human): Leahy, Axel & Hendrickson (1992) PDB1CD8
C1 (human): Weiss, Fauser & Engel (1986) J. Mol. Biol. 189, 573, Cooper (1985) Adv. Immunol. 37, 151
Membrane Attack Complex (human): Tschopp (1984) J. Biol. Chem. 259, 7587
Von Willebrand Factor (human): Fowler, Fretto, Hamilton, Erickson & McKee (1985) J.Clin. Invest. 76, 1491
Tissue Factor (human): Muller & DeVos (1994) PDB1HFT
Factor VII (human): modeled after Factor X
Factor X (human): Tulinsky & Padmanabhan (1993) PDB1HCG
Thrombin (cow): Vitali & Edwards (1993) PDB1HRT
Fibrin (cow): Weisel, Stauffacher, Bullitt & Cohen (1985) Science 230, 1388

Chapter 6

Histamine: idealized coordinates
Histidine Decarboxylase (bacteria): Gallagher, Rozwarski, Ernst & Hackert (1992) PDB1PYA
Prostaglandin H$_2$ Synthase (ram): Picot, Loll & Garavito (1994) PDB1PRH
Oxytocin (human): Husain, Blundell, Wood, Tickle, Cooper & Pitts (1987) PDB1XY1
Endorphin (human): idealized alpha helix
Human Growth Hormone: DeVos, Ultsch & Kossiakoff (1993) PDB3HHR
Epidermal Growth Factor (mouse): Kohda & Inagaki (1992) PDB1EPI
Interleukin-8 (human): Clore & Gronenborn (1990) PDB1IL8
Interferon- γ (cow): Samudzi & Rubin (1993) PDB1RFB
Insulin (human): Hua, Shoelson, Kochoyan & Weiss (1992) PDB1HIU
Glucagon (pig): Blundell, Sasaki, Dockeril & Tickle (1977) PDB1GCN
Thyroxine: idealized coordinates
Transthyretin (human): Wojtczak, Luft & Cody (1991) PDB1THA
Estradiol and Testosterone: idealized coordinates
Renin (human): Gruetter, Rahuel & Priestle (1991) PDB1RNE
Growth Hormone Receptor (human): DeVos, Ultsch & Kossiakoff (1993) PDB3HHR
G-Linked Hormone Receptor: Findlay & Eliopoulos (1990) TIBS 11, 492
G protein: Coleman, Berghuis, Lee, Linder, Gilman & Sprang (1994) Science 265, 1405
Calmodulin (rat): Babu, Bugg & Cook (1988) PDB3CLN
Rhodopsin (human): Findlay & Eliopoulos (1990) TIBS 11, 492
Crystallins (cow): White, Driessen, Slingsley, Moss & Lindley (1989) PDB2GCR, Bax, Lapatto, Nalini, Driessen, Lindley, Mahadevan, Blundell & Slingsley (1992) PDB2BB2
Sodium–Potassium Pump (pig): Skriver, Kavens, Hebert & Maunsbach (1992) J. Struct. Biol. 108, 176
Neurotransmitters: idealized coordinates
Acetylcholine Receptor (torpedo ray): Unwin (1993) J. Mol. Biol. 229, 1101

Index

Abductin, 102
ABO blood group, 17–18
Acetaminophen, 119
Acetylcholine esterase, 163
Acetylcholine receptor, 110, 163
Acetylcholine, 161
Acid, taste of, 158
Aconitase, 72
Aconitine, 161
Actin, 5, 90–92, 105–107
Actinidin, 58
Activators, 46–47
Active site, 20, 21
Adenine, 15, 43; in NAD, 73
Adenylate kinase, 79
Adrenal gland, 149
Aerobic exercise, 74–75
Affinity maturation, 126
Alanine, 10–11
Alcohol dehydrogenase, 33–34
Alcohol, 162; breakdown of, 33–34, 120;
    synthesis of, 69–70
Aldolase, 69
Allergies, 127
Allosteric proteins, 23–25, 75–77
Amino acids, 9–15; essential, 14–15; synthesis
    of, 25–29; and the genetic code, 47–49;
    taste of, 158
Aminoacyl-tRNA synthetases, 48–49
Aminopeptidase, 61
Ammonia, transport of, 12, 31–33
Amphetamines, 162
Amylase, 61–62

Anaerobic exercise, 70, 74–75
Anaphylactic shock, 113
Anchoring fibril, 96–98
Angiotensin, 150–151
Antibiotics, 29–30, 50
Antibodies, 5, 124–127; as antidotes, 111; in
    anaphylactic shock, 113
Anticoagulants, 135
Anticodon, 48–49
Antidiuretic hormone, 144
Antidotes, 90–92, 110–111
Antihistamines, 127, 143
Antihypertensives, 151
Antioxidants, 120
Apatite, 99
Arginine, 10–11
Asparagine, 10, 12
Aspartame, 158
Aspartate aminotransferase, 27–29
Aspartate carbamoyltransferase, 23–25
Aspartate, 10–12; in proteinases,
    59–60
Aspirin, 134, 144
Atherosclerosis, 65
Atoms, size of, 6–7
ATP synthase, 74–75
ATP, 67–68; in blood clotting (ADP), 132; in
    muscle contraction, 105; powering
    sodium-potassium pump, 159–160;
    synthesis of, 68–75; transfer of phos-
    phate, 79–80;
Avidity, 126
AZT (azidothymidine), 16

Index

Bacteria, defense against, 121–122, 130–131; limited genetic information of, 37; nitrogen fixing, 58; and porin, 86–87; toxins of, 113–116
Banded fibril, 96–97
Barbiturates, 162
Base pairing, 40–41
Bases, nucleic acid, 15
Bee venom, 113
Bicarbonate, 22
Bilayer, lipid, 5, 82–85
Biotechnology, 35, 42, 66–67
Birth control pills, 150
Blindness, color, 155; smell, 157
Blood, 4–5; ABO type, 17–18; clotting, 131–137; vessel dilation, 141
Bone, 99
Botulism toxin, 116

C1 (complement), 130
Calcium pump, 88, 90
Calcium, 87–88, 99, 153; amount in cells, 90
Calmodulin, 153
Calories, 67
Cancer, causes of, 38–39, 146–147; chemotherapy, 30–31, 92, 147
Carbohydrate, 16–18, 99; digestion of, 61–63
Carbon dioxide, 22
Carbonic anhydrase, 22–23
Carboxypeptidase, 60–61
Cardiotoxins, 110–111
Carotene, 154
Cascade, blood clotting, 134–135; complement, 130–131
Catalase, 120–121
Cataracts, 157
Cathepsins, 58
CD (cluster of differentiation) proteins, 129–130
Cells, 3–5; division of, 92
Cellulose, 16
Chaperones, molecular, 50–52
Chloramphenicol, 50
Cholera toxin, 114
Cholesterol, 83–84, 64–65, 149
Chromosomes, 39

Chylomicrons, 64
Chymosin, 59–60
Chymotrypsin, 21–22, 57
Cilia, 107
Cirrhosis, 117
Citrate synthase, 72
Citric acid cycle, 70–72
Codon, 47–49
Coenzyme Q, 73–74
Collagen, 95–99; and snake venom, 110
Color vision, 154–155
Complement, 130–131
Connective tissue, molecules of, 95–104
Connexons, 87–88
Cortisol (hydrocortisone), 144, 149
Crosslinks, by cysteine, 12; formation of, 51–52, 119; in proteins, 95, 102
Crystallins, 155–157
Curare, 163
Cyclic AMP, 153
Cyclooxygenase, 134, 144
Cysteine, 10, 12; crosslinks, 12, 51; in glutathione, 118–119; metal binding of, 123; in proteinases, 58
Cytochrome c, 73–74
Cytochrome oxidase, 73–74
Cytochrome p450, 117–118
Cytochrome reductase, 73–74
Cytosine, 15

DDI (dideoxyinosine), 16
Deoxyribonuclease, 65–66
Deoxyribonucleic acid, see DNA
Depressants, 162
Detoxification, 116–124
Diabetes mellitus, 148
Diazepam (Valium), 162
Digestion, 56–64; by venom, 110
Digitoxin/digoxin, 160
Dihydrofolate reductase, 29–30
Dipeptidyl peptidase, 61
Diphtheria toxin, 113–114
Disaccharidases, 62–63
Dismutation, 120
Disulfide bond formation protein, 51
Disulfide, see Crosslinks

DNA, 15–16, 36–38; fingerprinting, 42; synthesis of, 40–42
DNA-binding proteins, 45–47
DNA gyrase, 39–40
DNA polymerase, 40–42
DNA topoisomerases, 39–40
Dopamine, 162
Drug discovery, 121–122
Drugs, see Antibiotics, Antihistamines, Antihypertensives, Cancer chemotherapy, and individual listings
Dynein, 106–107

Elastase, 57
Elastin, 100–102
Electron transport proteins, 72–74
Embryonic development, and fibronectin, 104
Endorphins, 145
Energy, 67
Enkephalins, 145
Enolase, 69
Enterotoxin, 114–115
Enzymes, 5, 20; active site of, 20–21; allosteric, 23–25; metals and, 22–23; see also individual listings
Error rate, see Fidelity
Escherichia coli, 114
Essential amino acids, 14–15
Estrogens, 149–150
Evolution, of mitochondria, 49–50; of proteins, 29, 37, 57–58
Eye color, 104
Eye lens, 155–157

Factors, blood clotting, 134–135
Fat, 63–64, 68
Fermentation, 69–70
Ferritin, 77
Fibrin/fibrinogen, 5, 135–137
Fibronectin, 102–104
Fidelity, DNA polymerase, 41–42; RNA polymerase, 45; ribosome, 49
Fingernails, 94–95
Folic acid, 29–30
Free radicals, 120

Fructose, 17, 34–35
Fumarase, 72

G-proteins, 114, 152–153
GABA (γ–aminobutyric acid), 161
Galactose, 17–18
Gall stones, 84
Gap junction, 87
Gated channels, 152
Gelatin, 99
Genes, 43
Genetic code, 47–48
Genetic information, 36–38
Glands, see Adrenal, Hypothalamus, Pituitary, Thymus, Thyroid
Glucagon, 147–148
Glucoamylase, 62
Glucose, 16–17, 34–35
Glucose-6-phosphate isomerase, 69
Glutamate, 10–12, as neurotransmitter, 161–162
Glutamate synthase, 31–33
Glutamine, 10, 12; and transport of ammonia, 31
Glutathione S-transferase, 118
Glutathione, 118–119
Glyceraldehyde-3-phosphate dehydrogenase, 69
Glycine, 10, 13; in collagen, 95; as neurotransmitter, 161–162
Glycogen, 16, 61–63, 68
Glycolysis, 68–70
Glycosaminoglycans, 99–101
Goiters, 149
GP120, 129–130
Growth factors, 146–147
Growth hormone receptor, 152
Growth hormone, 145
Guanine, 15

Hallucinogens, 162
Hair, color, 104; keratin, 94–95
HDL (high density lipoprotein), 65
Healing, 131–137
Heme, 75–76

Index

Hemlock, 163
Hemoglobin, 75–77; leghemoglobin, 33
Hemophilia, 135
Heredity, 37
Hexokinase, 69
Hirudin, 135
Histamine, 113, 127, 142–143
Histidine, 10, 13
Histidine decarboxylase, 142–143
Histones, 39
HIV, 16, 129–130
HIV proteinase, 59–60
Hormones, 140–153; synthesis of, 150–151;
    transport of, 149
Hormone receptors, 151–153
Hornet venom, 143
Hyaluronic acid, 99–100
Hydocortisone (cortisol), 144, 149
Hydrogen ions, in ATP synthesis, 74–75
Hydrogen peroxide, 120–121
Hydroxyproline, 14; in collagen, 95–96
Hypothalamus, 144–145

Ibuprofin, 144
Inflammation, 142–143, 149
Immune system, 124–131
Immunity, 114–116, 127
Information, genetic, 36–38
Inorganic pyrophosphatase, 79
Insulin, 147–148
Interferons, 146–147
Interleukins, 146–147
Intermediate filaments, 92, 94–95
Introns, 44
Iodine, 148–149
Iron, transport and storage of, 77–78; in
    enzymes, 33, 75–76, 120
Isocitrate dehydrogenase, 72
Isoleucine, 10, 13

Keratin, 94–95
α–Ketoglutarate dehydrogenase, 71–72
Kinesin, 106–107

Lactase, 62–63
Lactic acid, 70

Lactose, 17; intolerance, 62–63
Laminin, 102–103
Latex, 102
LDL (low density lipoprotein), 64–65
Leeches, 135
Legumes, and nitrogen fixation, 33
Leucine, 10, 13; zipper, 46–47
Light, sensing of, 154–155
Lipases, 63–64; in venom, 110
Lipid, 5, 82–85; digestion and transport of,
    63–65
Lipoprotein, serum, 5, 64–65
Lipoprotein lipase, 64
Liver, and detoxification, 117
Lockjaw, 116
LSD (lysergic acid diethylamide), 162
Lymphocytes, 124–130
Lysine, 10–11
Lysosomes, 58, 121–122

α–Macroglobin, 122–123
Malate dehydrogenase, 72
Maltase-glucoamylase, 62–63
Melanin, 104–105
Membranes, 5, 82–85; proteins and, 84–85
Membrane attack complex, 130–131
Mescaline, 162
Melittin, 111, 113
Memory, 164
Messenger RNA, 42–44
Metals, in enzymes, 22–23; storage of, 123–
    124
Metallothionein, 123–124
Methionine, 10, 12; as start of protein
    synthesis, 49
MHC (major histocompatibility complex),
    127–129
Microscopy, 3
Microtubules, 92–93
Minerals, 99
Mitochondria 73–74; evolution of, 49–50; and
    porin, 85
Molecular chaperones, 50–52
Molecules, 1–2; names of, 8; size of, 3, 5–6;
    structure of, 8–18
Molybdenum, 33
Morphine, 145

Motor proteins, 105–107
MSG (monosodium glutamate), 158
Muscle contraction, 90, 105–107
Mutations, 38–39, 41–42
Myosin, 105–107

NAD (nicotinamide adenine dinucleotide), 72–
    73; and alcohol dehydrogenase, 33–34;
    in energy production, 69–73
NADH dehydrogenase complex, 73–74
Nerve transmission, 140, 159–164
Nettles, 143
Neurotoxins, 110–112
Neurotransmitters, 161–163; in bee venom, 113
Niacin, 73
Nitric oxide, 141
Nitrogenase, 33
Nitroglycerin, 141
NMDA (N-methyl-D-aspartate) receptor, 164
NMR (nuclear magnetic resonance) spectros-
    copy, 7
Norepinephrine, 162
NSAIDS (non-steroidal antiinflammatory
    drugs), 144
Nuclear pore, 88–89
Nucleases, 65–67
Nucleic acids, 15–16; digestion of, 65–67; see
    also DNA and RNA
Nucleoside diphosphate kinase, 79–80
Nucleosome, 39
Nucleotides, 15–16; synthesis of, 23–25, 30–
    31, 79–80

Odorant receptors, 157
Oncogenes, 146–147
Opsins, 154–155
Oxygen, 70, 75–77, 117, 119–120
Oxytocin, 144–145

p53 tumor suppressor, 38–39
Pain, 143–145
Pancreas, and blood sugar, 147
Papain, 58
PCR (polymerase chain reaction), 42
Penicillin, 122

Pepsin, 59
Peptidases, 60–61
Peroxidase, 144
Peroxisomes, 120
Phenylalanine, 10, 13–14
Phosphate, in ATP, 68, 79–80; in bone, 99
Phosphofructokinase, 69
Phosphoglycerate kinase, 69
Phosphoglycerate mutase, 69
Phospholipase, 64, 143–144
Phospholipids, 83–84, 143–144
Pituitary gland, 144–145
Plasmin, 137
Platelet, 4, 132, 134
Poisons, 110; of actin, 90–92; of nerve
    transmission, 160–163; of protein
    synthesis, 45, 50
Polymerases, see DNA polymerase and RNA
    polymerase
Polysaccharides, see carbohydrates
Porin, 85–87
Potassium, and sodium-potassium pump,
    159–160; taste of, 158
Proenzymes, 56
Progestins, 149
Proline, 10, 14; in collagen, 95–96; in protein
    folding, 51–52
Proline cis-trans isomerase, 51–52
Prostaglandins, 143–144
Prostaglandin synthase, 144
Proteinases, 56–60; of complement, 130–131;
    protection against, 122–123
Proteins, 9–15; destruction of obsolete, 52–
    53, digestion of, 56–61; folding, 9–11,
    22, 50–52; lifespan of, 52; membrane-
    bound, 84–85; synthesis of, 36–53, 114;
    transport of, 88; see also Enzymes,
    Hormones, Motor proteins, Receptor
    proteins, and individual listings
Proteoglycans, 99–101
Proteosome, 53
Pyrophosphate, 79
Pyruvate dehydrogenase complex, 70–71
Pyruvate kinase, 69

Receptor proteins, 151–153
Red blood cell, 4, 75–76

Index

Regulation, of enzymes, 23–25
Renin, 150–151
Repressor proteins, 46–47
Resilin, 102
Restriction endonucleases, 66–67
Retinal, 154–155
Rhodopsin, 154–155
Ribonuclease, 65–66
Ribosomes, 5, 49–50
Ribozymes, 44
RNA, 15–16, 44–45, 88; see also Ribosomes, Messenger RNA and Transfer RNA
RNA polymerase, 44–45; and DNA-binding proteins, 45–47
Rubber, 100–102

Saccharine, 158
Saliva, 61
Satellite sequences, 37
Saturated lipids, 84
Scorpion venom, 111–112
Scurvy, 96
Sequence-specific DNA-binding proteins, 46–47
Senses, 154–158
Serine, 10, 14; in proteinases, 21, 56–58
Serotonin, 161–162
Serpins, 57–58
Serum, blood, 5
Serum lipoproteins, 5, 64–65
Sex hormones, 149–150
Sex-linked genes, 155
Signal sequences, 88
Size, enzymes, 21–22
Skin, color, 104–105; keratin, 94–95
snRNP (small nuclear ribonucleoproteins), 44
Smell, 157
Snake venom, 110–111
Sodium, in nerve transmission, 159–160; taste of, 158
Sperm, 17, 107
Spider venom, 111–112
Splicing, of messenger RNA, 44
Starch, 16, 61–63
Steroid, analgesics, 144; hormones, 149–150
Streptomycin, 50

Strychnine, 162
Substrate channeling, 26–27
Succinate dehydrogenase, 72
Succinyl-CoA Synthetase, 72
Sucrase-isomaltase, 62–63
Sucrose, 17
Sugars, simple, 16–17; storage of, 62; taste of, 158
Sulfa drugs, 29–30
Superoxide dismutase, 119–120
Synapse, 162

Tanning, 104–105
Taste receptors, 157–158
Taxol, 92
T-cells, 129–130
T-cell receptor, 128
Tenascin, 103
Testosterone, 149–150
Tetanus toxin, 116
Tetracycline, 50
Thought, 159–164
Threonine, 10, 14
Thrombin, 134–145
Thromboxane A_2, 134
Thymidylate synthase, 30–31
Thymine, 15; synthesis of, 30–31
Thymus gland, 125–126
Thyroid gland, 148–149
Thyroxine, 148–149
Tissue factor, 134
Topoisomerases, 39–40
Toxins, 109–116
Transfer RNA, 47–49
Transferrin, 77–78
Transposable sequences, 37
Transthyretin, 148–149
Traveler's diarrhea, 114
Triglyceride, 63–64
Trimethoprim, 29
Triose phosphate isomerase, 69
Trypsin, 57–58
Trypsin inhibitor, 58
Tryptophan, 10, 13–14; synthesis of, 25–27
Tryptophan synthase, 25–27

Tyrosinase, 104
Tyrosine, 10, 13–14

Ubiquitin, 51–52
Ultraviolet light, 104
Unsaturated lipids, 84
Uracil, 15

Vaccination, 114–116, 127
Valine, 10, 13
Vanadium, 33
Vasopressin, 144
Venom, 109–113
Vitamins, 27–29; A, 154; B, 27–29; C, 96,
    120; E, 120; K, 135; niacin, 73
Viruses, limited genetic information of, 37–38;
    protection against, 65–66, 127–128

Voltage-gated ion channels, 160–161
von Willebrand factor, 132–134

Warfarin, 135
Wasp venom, 113
Water, and enzymes, 22, 55; and protein
    folding, 9
White blood cells, 4–5, 120–134
Wound healing, 131–137

X-ray crystallography, 3–7
D-Xylose isomerase, 34–35

Zinc, in enzymes, 22–23, 60; finger, 46–47
Zymogens, 56